International Workshop on Electronic Spectroscopy for Gas-Phase Molecules and Solid Surfaces 2009

(IWES 2009)

Satellite Meeting of ICESS-11
Journal of Physics: Conference Series Volume 235

Matsushima, Japan
12-15 October 2009

ISBN: 978-1-61738-658-9
ISSN: 1742-6588

Printed from e-media with permission by:

Curran Associates, Inc.
57 Morehouse Lane
Red Hook, NY 12571

CURRAN ASSOCIATES INC. proceedings .com

Some format issues inherent in the e-media version may also appear in this print version.

Printed by Curran Associates, Inc. (2010)

Institute of Physics
Dirac House, Temple Back
Bristol BS1 6BE UK

Phone: 44 1 17 929 7481
Fax: 44 1 17 920 0979

techtracking@iop.org

Additional copies of this publication are available from:

Curran Associates, Inc.
57 Morehouse Lane
Red Hook, NY 12571 USA
Phone: 845-758-0400
Fax: 845-758-2634
Email: curran@proceedings.com
Web: www.proceedings.com

International Workshop on Electronic Spectroscopy for Gas-Phase Molecules and Solid Surfaces 2009

(IWES 2009)

Satellite Meeting of ICESS-11
Journal of Physics: Conference Series Volume 235

Matsushima, Japan
12-15 October 2009

ISBN: 978-1-61738-658-9
ISSN: 1742-6588

IOP Institute of Physics

TABLE OF CONTENTS

012001 Using Photoelectron Diffraction to Determine Complex Molecular Adsorption Structures 1
D P Woodruff

012002 Theoretical Studies of Autoionization Dynamics of High Rydberg States 10
Yi-Hsieh Wang, H Mineo, S D Chao, Y Teranishi, S H Lin, H L Selzle, H J Neusser, E W Schlag

012003 Scanning Tunneling Microscopy and Inelastic Electron Tunneling Spectroscopy Studies of Methyl Isocyanide Adsorbed on Pt(111) 20
Satoshi Katano, Yousoo Kim, Masafumi Hori, Michael Trenary, Maki Kawai

012004 Dynamics of Single Or Double Ionization of Small Systems from Coincidence Electron Impact Experiments 27
A Lahmam-Bennani, E M Staicu Casagrande, A Naja

012005 Spin-polarised Stokes Parameters of Open and Closed Shell Transitions and Their Resonances in Zinc Atoms 33
J F Williams, L Pravica, D Cvejanovic, S Napier, S Samarin, M Piwinski

012006 Photoelectron Imaging in Pump-probe Experiments Combining Synchrotron and Laser Radiation 45
P O'Keeffe, P Bolognesi, R Richter, A Moise, E Ovcharenko, L Pravica, R Sergo, L Stebel, G Cautero, L Avaldi

012007 The Valence Shell Excitations for Some Noble Atoms Studied by Fast Electron Impact 53
Lin-Fan Zhu

012008 Silicon Thermal Oxidation and Its Thermal Desorption Investigated by Si 2p Core-level Photoemission 62
Y Enta, H Nakazawa, S Sato, H Kato, Y Sakisaka

012009 Imaging Magnetic Atoms by Photoelectron Diffraction 68
Takahisa Ishiyama, Masahiko Wada, Kouta Ito, Takehisa Konishi, Takashi Fujikawa

012010 X-ray Photoelectron Diffraction Study of Bias-treatment for the Growth of 1-inch-diameter Hetero-epitaxial Diamond (001) Thick Films 74
S Kono, H Kawata, T Goto, T Abukawa, K Chigira, K Ooyama, T Kotaki, A Sawabe

012011 Excitation Power Dependence of Lifetime of Photoluminescence from Gold Film in the Kretschmann Geometry 80
Tomonori Sanbongi, Satoshi Katano, Jamal Uddin Ahamed, Yoichi Uehara, Sukekatsu Ushioda

012012 Collision-energy-resolved Metastable Impact Electron Spectroscopy of Thiophene and 2,2'-Bithiophene: Stereodynamics in Gas Phase and Adsorbed Layer on Ag(110) 85
S Kudo, N Kishimoto, K Ohno

012013 (e, 2e) Processes on Ne, Ar and Xe Targets 93
G Purohit, Vinod Patidar, K K Sud

012014 Two-dimensional Penning Ionization Electron Spectroscopy of CH_3I and CH_2I_2 by $He^*(2^3s)$ Metastable Atoms 101
Naoki Kishimoto, Andriy Borodin, Takuya Horio, Masakazu Yamazaki, Koichi Ohno

012015 Radiative Charge Transfer and Interatomic Coulombic Decay Following Direct Double Photoionization of Neon Dimers 109
I Higuchi, T Ouchi, K Sakai, H Fukuzawa, X-J Liu, K Ueda, H Iwayama, K Nagaya, M Yao, D Ding, D Zhang, Y Tamenori, N Saito

012016 Vibrationally Resolved Resonant X-ray Emission Spectra of Diatomic Molecules 116
M Oura, O Takahashi, T Gejo, T Tokushima, Y Horikawa, Y Senba, H Ohashi, S Shin

012017 Hydrogen Bonding Interaction of Small Acetaldehyde Clusters Studied with Core-electron Excitation Spectroscopy in the Oxygen K-edge Region 127
K Tabayashi, M Chohda, T Yamanaka, Y Tsutsumi, O Takahashi, H Yoshida, M Taniguchi

012018 X-ray Absorption Spectra of SiF_4 and $Si(CH_3)_4$ in the Si K-shell Excitation Region 137
Osamu Takahashi, Isao H Suzuki, Yutaro Kono, Akihiro Ikeda, Takao Ouchi, Kiyoshi Ueda, Itaru Higuchi, Yusuke Tamenori, Shin-Ichi Nagaoka

012019 Formation of the Energetic Doubly Charged Ne Ion by Irradiation of Large Neon Clusters Using Intense EUV-FEL Pulses at 52 NM 145
K Nagaya, A Sugishima, H Iwayama, H Murakami, M Yao, H Fukuzawa, X-J Liu, K Motomura, K Ueda, N Saito, A Rudenko, M Kurka, K-U Kühne, J Ullrich, L Foucar, A Czasch, R Dörner, R Feifel, M Nagasono, A Higashiya, T Togashi, M Yabashi, T Ishikawa, H Kimura, H Ohashi

012020 Vibrational Spectra and Geometry Relaxation in Core-electronic Processes of N_2O and CO_2 150
Masahiro Ehara, Kiyoshi Ueda

Author Index

International Workshop on Electronic Spectroscopy for Gas-phase Molecules and Solid Surfaces (IWES2009) was held in Matsushima, Japan for the period 12–15 October 2009. IWES was a satellite meeting of the Eleventh International Conference on Electronic Spectroscopy and Structure (ICESS-11) held in Nara, Japan, for the period 6–10 October 2009. The workshop covered a broad range of experimental and theoretical hot topics in electronic spectroscopy for gas-phase molecules and solid surfaces. About 90 scientists from 10 countries attended the workshop and discussed intensively future prospects of the field. A total of 19 invited talks, 17 contributed talks, and 53 contributed posters covered recent advances in these topics involving excitation, ionization, break-up, etc. of a wide range of targets by photons and charged particles. These proceedings present papers on 7 of the invited papers and 13 of contributed papers.

One of the reasons for this workshop was that one of the co-chairs, Professor Shozo Kono, was retiring - he retired from Tohoku University in March 2010. Many of his friends attended the workshop, celebrating his scientific achievement. This proceedings volume is thus dedicated to Prof. Shozo Kono on the occasion of his retirement.

The workshop was hosted by Institute of Multidisciplinary Research for Advanced Materials, Tohoku University. The Organizers gratefully acknowledge the financial support of Morino Foundation, Tohoku Intelligent Cosmos, and Casio Foundation. We also thank steering committee members for their invaluable assistance in the smooth running of the workshop and international advisory board members for their invaluable input for making attractive scientific program.

Kiyoshi Ueda
Tohoku University
on behalf of organizing committee
http://res.tagen.tohoku.ac.jp/~iwes2009

Organizing committee

K. Ueda (Co-Chair) S. Kono (Co-Chair)
M. Takahashi (Secretary) T. Abukawa (Treasurer)
T. Komeda

Steering committee

M. Takahashi (Co-Chair) T. Abukawa (Co-Chair)
Y. Takakuwa M. Okunishi
N. Watanabe T. Takaoka
Y. Sainoo M. Yamazaki
H. Fukuzawa Y. Kino

N. Kishimoto

International advisory board

E. Bauer (USA)

W. Eberhardt (Germany)

C. S. Fadley (USA)

A. Ichimiya (Japan)

H. Kono (Japan)

P. Morin (France)

H. Ohta (Japan)

M.-N. Piancastelli (Sweden)

K. Prince (Italy)

J. Williams (Australia)

D.P. Woodruff (UK)

All papers published in this volume of *Journal of Physics: Conference Series* have been peer reviewed through processes administered by the proceedings Editors. Reviews were conducted by expert referees to the professional and scientific standards expected of a proceedings journal published by IOP Publishing.

International Workshop on Electronic Spectroscopy for Gas-phase Molecules and Solid Surfaces IOP Publishing
Journal of Physics: Conference Series **235** (2010) 012001 doi:10.1088/1742-6596/235/1/012001

Using photoelectron diffraction to determine complex molecular adsorption structures

D.P. Woodruff

Physics Department, University of Warwick, Coventry CV4 7AL, UK

email: d.p.woodruff@warwick.ac.uk

Abstract. Backscattering photoelectron diffraction, particularly in the energy-scan mode, is now an established technique for determining in a quantitative fashion the local structure of adsorbates on surfaces, and has been used successfully for ~100 adsorbate phases. The elemental and chemical-state specificity afforded by the characteristic core level photoelectron binding energies means that it has particular advantages for molecular adsorbates, as the local geometry of inequivalent atoms in the molecule can be determined in a largely independent fashion. On the other hand, polyatomic molecules present a general problem for *all* methods of surface structure determination in that a mismatch of intramolecular distances with interatomic distances on the substrate surface means that the atoms in the adsorbed molecule are generally in low-symmetry sites. The quantities measured experimentally then represent an incoherent sum of the properties of each structural domain that is inequivalent with respect to the substrate point group symmetry. This typically leads to greater ambiguity or precision in the structural solutions. The basic principles of the method are described and illustrated with a simple example involving molecule/substrate bonding through only one constituent atom (TiO_2(110)/H_2O). This example demonstrates the importance of obtaining *quantitative* local structural information. Further examples illustrate both the successes and the problems of this approach when applied to somewhat more complex molecular adsorbates.

1. Introduction

A detailed knowledge of the structure of adsorbed molecules on surfaces is the starting point to understanding the electronic structure, the adsorbate-substrate bonding, and the potential chemical properties. Key aspects are the adsorption site, the extent to which the molecule and/or the underlying surface are structurally modified by the bonding, and the interatomic bondlengths, particularly between the molecule and the surface, but also within the molecule. Heterogeneous catalysis exploits the change in energetics of a reaction between two species when at least one of them is bonded to an 'active' surface, so these structural questions underpin any attempt to gain a fundamental understanding of this crucially important aspect of chemistry.

Developments in the techniques of surface science have led to the availability of a range of methods to determine the structure of single crystal surfaces with and without adsorbate species. Conventional diffraction techniques, and low energy electron diffraction (LEED) in particular, have contributed the largest number of structural solutions[1], but these methods rely on long-range order of the whole surface. For many molecular adsorbate systems, and particularly those involving coadsorption of two or more species, a situation of particular relevance to understanding surface reactions and the role of surface promoters and poisons, the adsorbates lack this long-range periodicity. To investigate such surfaces one requires a *local* structural probe, and the most successful one has been photoelectron diffraction. This technique exploits the same coherent interference of elastic scattering paths as LEED, but uses a local electron source, namely photoemission from a core level of a constituent atom of the adsorbate species.

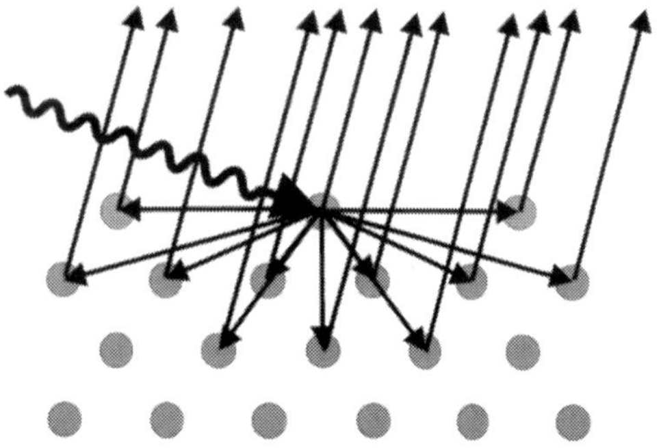

Fig. 1 Schematic of the electron scattering paths in photoelectron diffraction from adsorbed atoms

Fig. 1 illustrates the underlying principle. The photoelectron wavefield emerges from the adsorbate atom (shown in Fig. 1 as an isolated atom, but this could equally well be an atom within an adsorbed molecule) and the detected signal outside the surface comprises a coherent sum of the directly emitted wave and components of the same wavefield elastically scattered from the atoms surrounding the emitter, including, in particular, the underlying substrate atoms. Modulations in the detected photoemission intensity are produced as a function of detection angle (which influences the scattering path lengths) or by changing the photoelectron wavelength by varying the photon, and thus the photoelectron, energy. As the phases of the scattering paths depend on the location of the emitter atom relative to its neighbours, the detected modulations are characteristic of the position of the emitter atom. In our investigations we exploit the scanned-energy mode of this technique, to which we have assigned the acronym PhD. The PhD modulation spectra are collected within an energy range of ~50-400 eV, for which elastic backscattering cross-sections are reasonably strong, and are measured at a range of emission angles to enlarge the data set and sample a larger range of electron momentum transfer space. Structure determination is then achieved by a trial and error approach based on a series of 'guessed' structures, that are each optimised by adjusting the structural parameters and comparing the results of multiple scattering simulations with the experimental results. This approach is essentially identical to that used in conventional LEED studies, and, indeed, in almost all surface structural techniques. The very first 'demonstration' experiments of photoelectron diffraction were performed in 1978[2,3,4], but since that time the method has been developed into a practical tool that has solved ~100 surface structures [5]

PhD has a number of important advantages for determining the local geometry of adsorbed molecules. Firstly, as noted above, it is intrinsically a local technique (resulting from the fact that the initial outgoing electron wave is some sum of spherical harmonics centred on the emitter atom). Secondly, it is element specific. The characteristic core-level binding energies of the different atomic species on the surface means that the PhD modulation spectra from each different elemental atom can be collected independently. Moreover, as most molecular adsorbates are comprised primarily of low atomic number elements (especially C, N, O and H), whereas most substrates of interest have significantly higher atomic number, the dominant elastic scattering contributions are generally from the substrate. This means that the location of each elemental atom on the surface can be determined in a fashion that is largely independent of the location of the other atoms within the molecule, greatly simplifying the structure determination. Intramolecular scattering can generally be added in a second stage of the analysis to refine the structure. Thirdly, PhD is chemical-state specific. If the absorbed species contains the same element in different local geometries, there will be a 'chemical shift' in the core level photoelectron binding energy; this effect is an important aspect of the technique of X-ray photoelectron spectroscopy (XPS) and, indeed, the origin of its original name of ESCA (electron spectroscopy for *chemical* analysis). Providing the associated chemical shift is large enough to be resolved, this means that the PhD data (and thus the local structure) can be separated for atoms of the same element in different structural environments. For molecular adsorbates this is most commonly exploited to distinguish atoms of the same element in different functional groups in an adsorbed molecule, for example C atoms in methyl (CH_3) and carboxylate (COO) species (e.g. in the acetate species[6]). However, it can also allow distinctions to be made between two identical species coadsorbed in different local sites on a surface, such as CO in atop, bridge and hollow sites.

While all these aspects are positive there are, of course, problems and limitations. These arise, particularly, because the technique works best when the emitter atom occupies a high-symmetry site on a surface. If the adsorption site has a symmetry that is lower than that of the point group symmetry of the substrate, then there will coexist on the surface multiple structural domains that are related by the 'missing' symmetry elements of the substrate. The measured PhD modulation spectra will then be an incoherent sum of the modulations from each of the symmetry-related domains, and the resulting modulation amplitude will be attenuated. Typically, in PhD, the strongest modulations are seen when a nearest-neighbour strongly-backscattering substrate atom lies directly 'behind' the emitter relative to the detection direction, in part for symmetry reasons, in part because elastic backscattering cross-sections commonly show a peak at 180° scattering angle. It is these strongly-modulated spectra that provide the most reliable basis for the associated structure determination and, indeed, the identification of the detection directions at which such spectra are measured provides a basis for initial identification of the most likely structural model. Low symmetry adsorption sites generally lead to weaker PhD modulations and thus to a more challenging process of structure determination. This is an intrinsic problem of many molecular adsorbate systems. If the molecule bonds to the surface through more that one constituent atom or, as is commonly the case for aromatic molecules, lies flat with its molecular plane essentially parallel to the surface, there is an inevitable mismatch of interatomic distances within the molecule and within the underlying surface. This means that at least some of the constituent atoms of the molecule are bound to occupy low symmetry sites (even if the molecule as a whole occupies a high-symmetry site).

In the remainder of this article a few examples of recent PhD studies are presented that illustrate both successes and challenges in molecular adsorbate studies on surfaces.

2. Molecular bonding through a single atom: $TiO_2(110)/H_2O$

When adsorbed molecules bond to a surface through a single constituent atom, any mismatch of intramolecular and interatomic surface bondlengths has no bearing on the way the molecule bonds to the surface, and one generally finds a high-symmetry adsorption site that can be solved relatively easily by the PhD technique. Fig. 2 shows results from such an investigation, notably of molecular water, H_2O, on $TiO_2(110)$[7,8].

Of course, in this system both the adsorbate and substrate contain O atoms, so it is important to be able to distinguish between emission from these two types of O emitter atoms. As may be seen from Fig. 2(a), which shows the XP spectra from the clean surface and after low-temperature water adsorption (and annealing to 230 K), there is a large chemical shift (3.5 eV) between the photoelectron binding energies of the water and oxide O 1s states. Fig. 2(b) shows a set of PhD modulation spectra recorded from the water O 1s component in 10 different geometries corresponding to different polar emission angles and three different azimuthal planes of incidence and photoelectron detection. The experimental data (full lines) are compared with the results of the multiple scattering simulations (dashed lines) for the best-fit structural model, shown in Fig. 2(c). Clearly the overall agreement is very good, reflected in a low associated reliability factor (R-factor) value of 0.15. One striking feature of the PhD spectra is the strong (~±40%) modulations that appear to have a single dominant period in the normal emission spectra (0° polar angle). This is the behaviour we may expect if the O atom of the water occupies a site directly atop a strongly-scattering substrate atom (such that normal emission corresponds to 180° scattering angle from this atom) and indeed the water is found to occupy a site atop the undercoordinated (5-fold coordinated) Ti atoms in the TiO_2 surface. The reduced sample temperature (~190 K) used in this study also helps to ensure that the PhD modulations are large, by reducing the Debye-Waller factor in the scattering process.

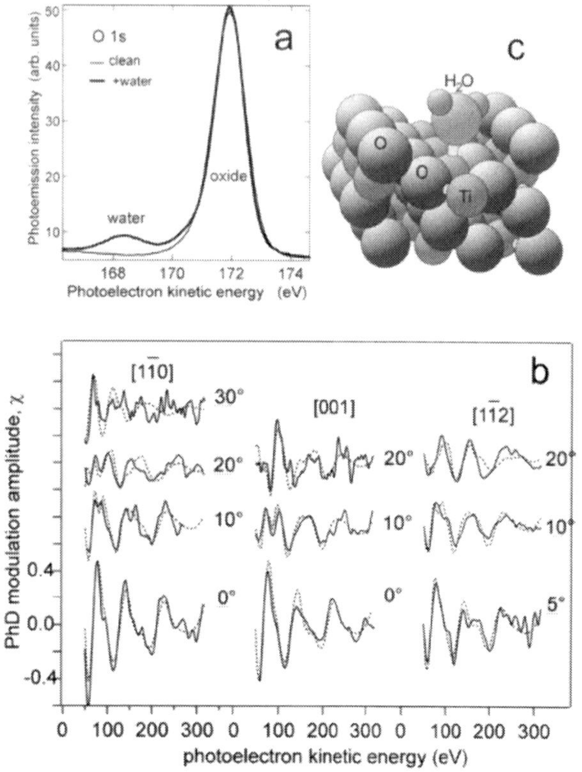

Fig. 2. Results of an investigation of the $TiO_2(110)/H_2O$ adsorption system[7,8]. (a) shows O 1s XP spectra from the clean and water-covered surface at a photon energy of 700 eV, (b) shows a set of experimental (full lines) and simulated (dashed lines) water O 1s PhD modulation spectra, (c) shows the structure associated with these best-fit simulations.

The adsorption site established in this way, atop an undercoordinated Ti atom, was actually identified in earlier STM (scanning tunnelling microscope) investigations of this surface[9,10], but a key feature of the PhD investigation is its ability to obtain *quantitative* information on the associated bondlengths. The TiO_2/H_2O system is of particular interest because of the possible application of titania as a catalyst for the photochemical production of hydrogen from water[11], but theoretical modelling of the interaction of water with $TiO_2(110)$ has proved challenging. In particular, while it is well-established experimentally that molecular water does not dissociate on a defect-free surface, most calculations in the past have concluded that such dissociation is facile. In this context the value of the $Ti\text{-}O_{water}$ bondlength of 2.21 ± 0.02 Å found in the PhD study appears to be highly relevant, because in theoretical studies that concluded that dissociation is facile the equivalent bondlength was reported to be very much longer, in the range 2.28-2.41 Å. In effect, it appears that molecular water is significantly more strongly bound to $TiO_2(110)$ than these calculations would suggest. This may well be the reason why the barrier to dissociation is calculated to be too low. Experimentally-determined adsorption bondlengths may thus provide a key test of theoretical calculations of this type.

Finally, it is perhaps worth noting that, in the context of PhD structure determination, this adsorption system is particularly simple, because H atoms are such weak electron scatterers that their role in the diffraction process can be safely ignored. As such, from the point of view of the data analysis (but not, of course, from the chemistry), the situation is equivalent to an investigation of a simple atomic (O) species alone.

3. Molecular bonding through two or more constituent atoms

The problem arising from low-symmetry emitter sites emerges when a molecule bonds to a surface through two or more of its constituent atoms, although the scale of the associated problem for PhD structure determination varies. A relatively simple situation arises, for example, in the case of simple carboxylic acids adsorbed on Cu(110). Interaction with the surface leads to deprotonation of the acid, and the resulting carboxylate bonds through the two O atoms in a symmetric fashion. This behaviour

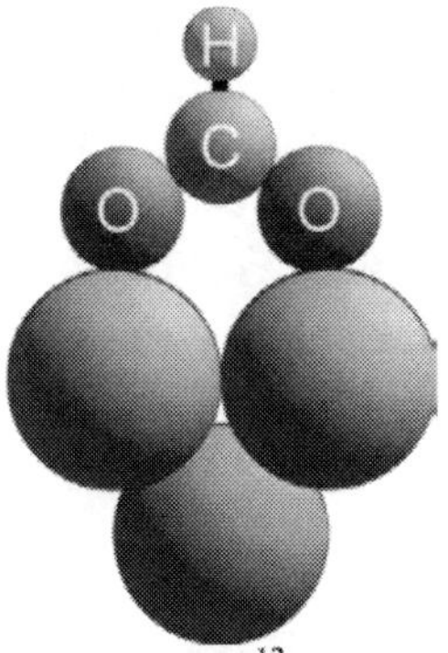

Fig. 3 Model of local geometry of formate on Cu(110)

has been confirmed for formate (HCOO)[12], acetate (CH_3COO)[6] and benzoate (C_6H_5COO)[13]. In each case the molecular plane is perpendicular to the surface and aligned in a close-packed $[1\bar{1}0]$ azimuth centred over a short-bridge site such that the two O atoms occupy symmetrically-equivalent off-atop sites (Fig. 3). The mismatch of the Cu-Cu distance in the surface (2.55 Å) and the O-O distance in a free formate species (2.22 Å) is modest, and indeed the O-C-O bond angle is somewhat flexible, so the two O atoms occupy sites within ~0.15 Å of the exact atop sites and the PhD data analysis does not present a problem.

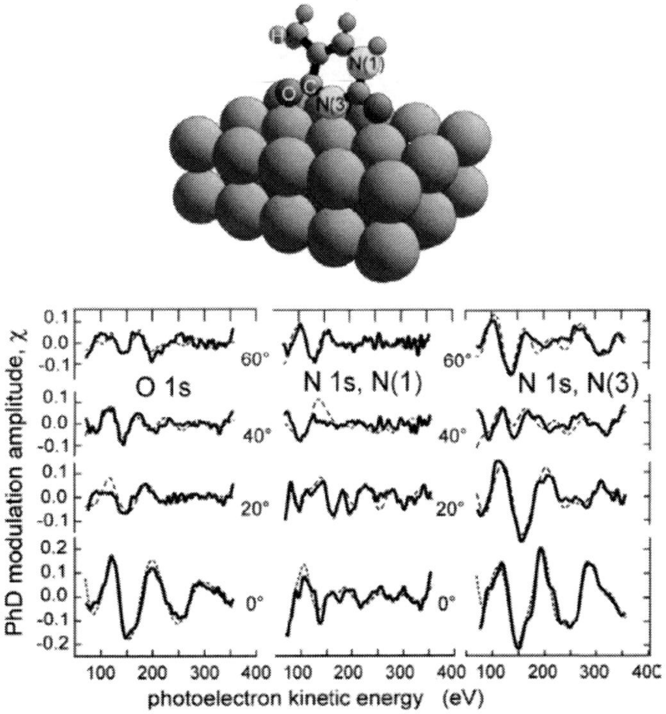

Figure 4. Optimised structural model and subset of the associated experimental (full lines) and simulated (dashed lines) PhD data from an investigation of the structure of thymine on Cu(110) at room temperature[14]. In the model the H atoms have been added for chemical consistency, but their locations were not determined.

A system that might have been expected to be significantly more challenging, illustrated in Fig. 4, is the adsorption of the nucleobase molecule thymine, on Cu(110). When deposited at room temperature XP spectra show two chemically-shifted N 1s peaks, attributed to dehydrogenation of one of the two NH species (notably the N(3) atom) through reaction with the surface to produce the molecular species shown in the schematic diagram of Fig. 4. The PhD structure determination[14] confirms this assignment, because the PhD spectrum from this N 1s component is closely similar to that of the two (equivalent) O atoms, showing a single dominant long period modulation at normal emission characteristic of a (near-) atop site. Thus, the molecule bonds to the surface through the N(3) atom and both the O atoms, all of which occupy off-atop sites. The amplitude of these modulations is about a factor of 2 weaker than those in the TiO_2/H_2O system, though, attributable to larger vibrational amplitudes (at room temperature) and the fact that the emitter atoms are displaced from the ideal atop sites. Interestingly, as also seen in fig. 4, the multiple-scattering simulations for this system also reproduce quite well the generally weak modulations from the N(1) N 1s emission component. As this N atom is rather far from the surface these modulations contain a significant contribution from intramolecular scattering.

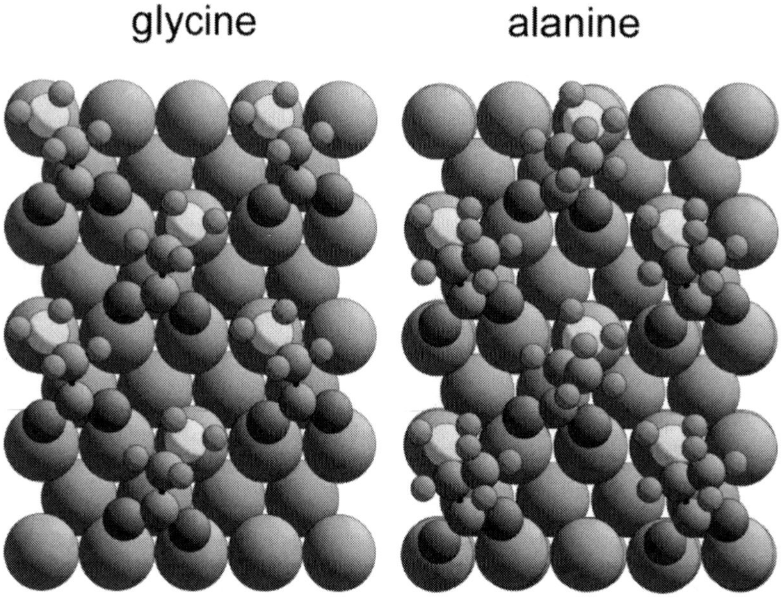

Fig. 5 Models of the (3x2) structures formed by glycinate and alaninate on Cu(110)

A rather different set of systems that have led to a lot of valuable information, but expose some constraints of the method, is concerned with simple amino acids on Cu surfaces and, in particular, on Cu(110). The fist study[15,16] was of the simplest amino acid, glycine (NH_2CH_2COOH) which, as with the simple carboxylic acids, loses its acid hydrogen on interaction with the surface to produce a surface glycinate (NH_2CH_2COO) species, while this was followed by an investigation[17] of the alaninate species produced by reaction with alanine ($NH_2CH_3C^*HCOOH$). A key difference is that in alanine one of the central H atoms is replaced by a methyl group, causing the associated (C*) carbon atom to become a chiral centre. Fig. 5 shows why this difference is of special interest in the surface phase formed on Cu(110). In the case of glycine, which is achiral as a free molecule, the glycinate species bond to the surface (through both carboxylate O atoms and the amino N atom) in such a way as to form chiral moieties that are 'left- and right-handed', mirror images of one another. These form an ordered (3x2) heterochiral structure. Interestingly, a single enantiomer of alanine also produces a (3x2) ordered phase, but in this case the molecules at the centre of the unit mesh cannot be mirror images of those at the corners, because this would involve a (forbidden) transition to the opposite enantiomer. The net effect is that the centred molecules in the unit mesh of the alanine structure must have a symmetrically distinct local adsorption structure from those at the corners.

The structure determination of both systems involved the measurement of the N 1s and O 1s PhD spectra in a range of detection directions. All showed clearly that the amino N atom lies close to atop a surface Cu atom, while the O atoms occupy two distinctly different sites that still appear to correspond to single coordination to surface Cu atoms but are up to ~1 Å off atop. The relatively low (2mm) symmetry of the (110) surface, and the mirror symmetry of the whole structure, mean that the number of structural domains over which the measurements average is only two (corresponding to the missing two-fold rotation axis of the surface phase), so quite strong modulations in the O 1s PhD spectra are seen around normal emission despite the large offset from atop. The structure determination provided an unambiguous measure of both the height of the O and N atoms above the surface, and of their lateral offsets from atop in the two principle azimuthal directions. The problem of how these individual atomic positions relate to one another in the whole molecule is not, however, entirely trivial. While it is clear that the O atom offset values in the [001] direction must have the same sign, the sign of the offset of the N atom offset in the $[1\bar{1}0]$ is less obvious. For example, for a specific molecule in Fig. 5, is the offset of the N atom from atop to the left or the right? Reasonable assumptions about the molecular conformation, relative to the known properties of the free glycine molecule, provide a rather clear answer to the question, yet this system does highlight a potential limitation of being able to determine the local site of the constituent atoms in a largely independent fashion. The strength offered by the simplicity of this approach can, in some cases, also become a limitation.

For the alanine system the problem of achieving a unique structure determination, based on the PhD data alone, becomes more serious. As discussed above, the four O atoms associated with the two inequivalent adsorbed alaninate species must all have different offset (low symmetry) sites, and obtaining a unique unconstrained solution proved impossible. The initial investigation clearly identified the basic geometry, but the final structural tests were performed on a previously optimised minimum energy structure obtained from density functional theory (DFT) calculations[18]. The very significant outcome was that while the lateral positions and substrate distortions found in the DFT results were consistent with the PhD data, the actual N-Cu and O-Cu bonding distances were not. Only after reducing these by ~0.1 Å was a good fit to the PhD data obtained. From the point of view of the chemisorption bonding, of course, these are the most important structural parameters.

A recent PhD study of furan, C_4H_4O, adsorbed on Pd(111)[19] provides an example of a more significant structural ambiguity. This investigation was greatly hampered by two problems. Firstly, the binding energy of the O 1s state is closely similar to that of the Pd $3p_{3/2}$ state so it proved impossible to extract reliable O 1s PhD spectra from the measurements. The difficulty in separating these components in the photoemission spectra is exacerbated by the fact that the Pd 3p spectra include shifted components associated with the surface layer[20], and as these are expected to be modified by adsorption, the clean surface spectra cannot be used a basis for separating the O and Pd components in the spectra from the furan-covered surface. The second problem is that the C 1s PhD spectra showed extremely weak modulations. In fact there are two chemically-shifted C 1s components, one associated with the two C atoms bonded to the O atom (the α-C atoms – see Fig. 6). The other with the two β-C atoms that are not bonded to the O atom. Modulation amplitudes of the α-C 1s PhD spectra were only ~±10% or less, but those of the β-C atoms were even weaker; under these circumstances it was felt that the signal-to-noise ratio of the β-C PhD modulation spectra were too low to allow a meaningful analysis, so the structure determination was constrained to the use of the α-C 1s PhD spectra alone. The fact that even these modulations are very weak clearly implies that the site of these C emitter atoms is substantially removed from a high symmetry site. An exhaustive search of possible structural models led to identification of the two basic structures shown in Fig. 6. In fact four optimal models were found, but the other two models only differ from the ones shown in Fig. 6 in being displaced from fcc hollows to hcp hollows. These two hollow sites (corresponding to positions directly above third and second layer substrate atoms, respectively, only differ with respect to these sub-surface layers, and for a planar molecule lying down on the surface the energetic difference in these pairs of bonding sites is expected (and found[21,22]) to be essentially negligible. All of these models share a common feature, namely that the β-C atoms are approximately 0.6 Å off atop sites with an associated C-Pd bondlength of 2.13±0.03 Å. As the structure is based on the PhD spectra from these β-C atoms, this level of structural identification is not surprising, but of course there are significant differences in the local bonding in the two distinct geometries shown in Fig. 6, so the insensitivity of the limited PhD data to their distinction does not imply that both structures are equally likely in practice. In fact, DFT calculations show that while both of these structures are stable, the 'hollow' geometry is energetically preferred by approximately 240 meV[21,22].

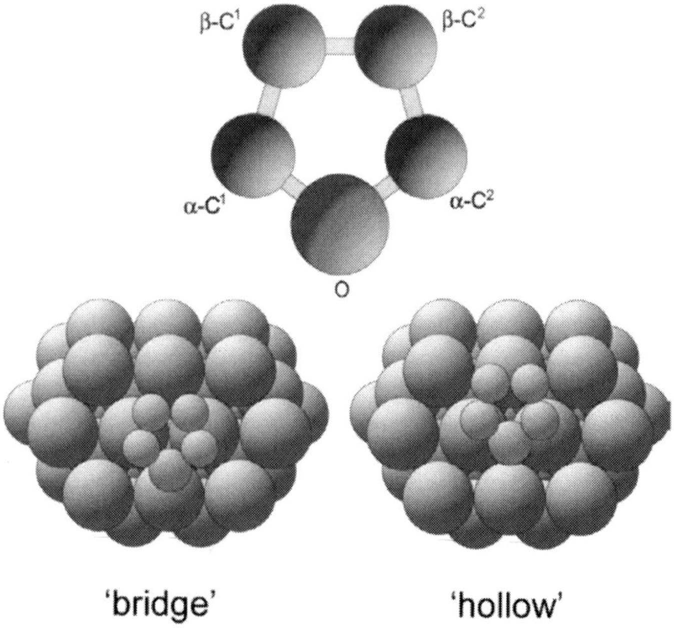

Fig. 6 Schematic diagram of the furan molecule (with the H atoms omitted) showing the labelling of the C atoms, and the two basic alternative models found for adsorption on Pd(111).

4. Conclusions

Three decades of development and exploitation of photoelectron diffraction as a means of obtaining quantitative structural information for adsorbates on surfaces have proved very fruitful, and in particular have led to a significant number of structural solutions of molecular adsorbates of increasing complexity. Such complexity does, however, lead to challenges and constraints in determining unique and precise structural solutions. These arise, particularly, due to constituent atoms commonly occupying low-symmetry sites relative to the underlying substrate. Despite these constraints, results continue to provide a necessary complement and challenge to the use of modern (DFT) total energy calculations of adsorbate structures, particularly raising questions regarding the accuracy of computed chemisorption bondlengths in such systems.

References

[1] Watson P R, Van Hove M A and Hermann K 2003 *NIST Surface Structure Database Ver. 5.0*, NIST Gaithersburg, MD

[2] Woodruff D P, Norman D, Holland B W, Smith N V, Farrell H H and Traum M M1978 *Phys. Rev. Lett.* **41** 1130
[3] Kono S, Goldberg S M, Hall N F T and Fadley C S 1978 *Phys. Rev. Lett.* **41** 1831
[4] Kevan S D, Rosenblatt D H, Denley D, Lu B-C and Shirley D A 1978 *Phys. Rev. Lett.* **41** 1565
[5] Woodruff D P 2007 *Surf. Sci. Rep.* **62** 1
[6] Weiss K-U, Dippel R, Schindler K-M, Gardner P, Fritzsche V, Bradshaw A M, Kilcoyne A L D and Woodruff D P 1992 *Phys. Rev .Lett.* **69** 3196
[7] Allegretti F, O'Brien S, Polcik M, Sayago D I and Woodruff D P 2005 *Phys. Rev. Lett.* **95** 226104
[8] Allegretti F, O'Brien S, Polcik M, Sayago D I and Woodruff D P 2006 *Surf. Sci.* **600** 1487
[9] Brookes I M, Muryn C A and Thornton G 2001 *Phys. Rev. Lett.* **87** 266103
[10] Schaub R, Thostrup P, Lopez N, Lægsgaard E, Stensgaard I, Nørskov J K and Besenbacher F 2001 *Phys. Rev. Lett.* **87** 266104
[11] Fujishima A and Honda K 1972 *Nature* **238** 37
[12] Woodruff D P, McConville C F, Kilcoyne A L D, Lindner Th, Somers J, Surman M, Paolucci G and Bradshaw A M 1988 *Surf. Sci.* **201** 228
[13] Pascal M, Lamont C L A, Kittel M, Hoeft J T, Terborg R, Polcik M, Kang J-H, Toomes R and Woodruff D P 2001 *Surf. Sci.* **492** 285
[14] Allegretti F, Polcik M and Woodruff D P 2007 *Surf. Sci.* **601** 3611
[15] Booth N A, Woodruff D P, Schaff O, Gieβel T, Lindsay R, Baumgartel P and Bradshaw A M 1998 *Surf. Sci.* **397** 258
[16] Kang J-H, Toomes R L, Polcik M, Kittel M, Hoeft J T, Efstathiou V, Woodruff D P and Bradshaw A M 2003 *J. Chem. Phys.* **118** 6059
[17] Sayago D I, Polcik M, Nisbet G, Lamont C L A and Woodruff D P 2005 *Surf. Sci.* **590** 76
[18] Rankin R B and Sholl D S 2005 *Surf. Sci.* **574** L1
[19] Knight M J, Allegretti F, Kröger E A, Polcik M, Lamont C L A and Woodruff D P 2008 *Surf. Sci.* **602**2524
[20] Sandell A, Beutler A, Jaworowski A, Wiklund M, Heister K, Nyholm R and Andersen J N 1998 *Surf. Sci.* **415** 411
[21] A. Loui, PhD. Thesis, University of California at Davis, 2005
[22] Bradley M K, Robinson J and Woodruff D P 2010 *Surf. Sci.* do:10.1016/j.susc.2010.02.021

Theoretical studies of autoionization dynamics of high Rydberg states

Yi-Hsieh Wang[1,2], H Mineo[1], S D Chao[1], Y Teranishi[3], S H Lin[2,3,4], H L Selzle[5], H J Neusser[5] and E W Schlag[5]

[1] Institute of Applied Mechanics and Center for Quantum Science and Engineering (CQSE), National Taiwan University, Taipei 106, Taiwan, ROC

[2] Research Center for Applied Sciences, Academia Sinica, Taipei 115, Taiwan, ROC

[3] Institute of Applied Chemistry, Institute of Molecular Science, Chiao-Tung University, Hsin-Chu, Taiwan, ROC

[4] Institute of Atomic and Molecular Sciences, Academia Sinica, Taipei 106, Taiwan, ROC

[5] Institut für Physikalische und Theoretische Chemie, Technische Universität München, Lichtenbergstr. 4, D-85748 Garching, Germany

E-mail: sdchao@iam.ntu.edu.tw

Abstract. In this paper we show how to employ the inverse Born-Oppenheimer approximation to establish a basis set to study zero kinetic energy (ZEKE) spectroscopy and the autoionization dynamics of the ZEKE Rydberg states. The calculations of vibrational and rotational autoionizations of a homonuclear diatomic molecule will be demonstrated as an example. The scaling laws for the autoionization rate constants are determined through nonlinear modeling of the calculated data.

1. Introduction

Zero-kinetic-energy (ZEKE) spectroscopy is a high resolution, rotationally resolved photoelectron spectroscopy (PES) to study the structures and dynamics of neutral and ionic molecules [1-4]. ZEKE spectroscopy has been increasingly used to determine the rovibronic energy levels of polyatomic molecular ions and to derive information on the dynamical and thermodynamical properties of molecular systems [4-9] and even the vibrational energy level structures of cations of biological molecules [10]. As in the case of conventional molecular spectroscopy, ab initio quantum chemistry calculation has become a powerful tool for the analysis of ZEKE spectra. This is especially true for the ionic species resulting from photoionization of bonding orbitals of the valence shell of stable and rigid molecules like hydrocarbons, which often exhibit a complex tunneling behavior necessitating careful studies of extended regions of the potential surfaces.

In a ZEKE experiment, a laser pulse will excite at least one electron to a high Rydberg state, which is just below the ionization threshold. After the excitation, these electrons would be detected by using a delayed ionizing field. Therefore, different from the conventional photoelectron spectroscopy (PES), which directly detects electrons in the continuum states, the electrons from the high Rydberg states in the ZEKE spectroscopy are observed [11-12]. For this reason, it is crucial to find a suitable representation of the high Rydberg state for studying the involved radiative or radiationless process. In this paper, we will present a theoretical model based on the inverse Born-Oppenheimer approximation

(IBOA) [13-16]. This model not only can provide a proper basis set for the system in high Rydberg states, but also for the dynamics of high Rydberg states, such as autoionization and collision [13, 17]. In section 2, we will present a mathematical derivation of IBOA and in section 3 we will show the numerical calculation of autoionization rate constants and the related scaling law.

2. Theory

2.1 Inverse Born-Oppenheimer approximation

Chao et al. [13-14] have developed a molecular theory of high Rydberg states using the inverse Born-Oppenheimer approximation (IBOA), which is a reversed version of the Born-Oppenheimer approximation (BOA). The basic idea is that we neglect the kinetic energy operator of the Rydberg electron in the Schrödinger equation, and obtain two separate differential equations. One is the ionic Schrödinger equation, whose Hamiltonian consists of the kinetic energy operators of core electrons and nuclei and the Coulomb potentials among all charged particles in a molecule. The other is the electronic Schrödinger equation of the Rydberg electron. With this separation, we can obtain the zeroth-order basis function for any atom or molecule in the high Rydberg state.

We now derive the basic equations. First, the Hamiltonian can be expressed as

$$\hat{H} = \hat{H}_{ion} + \hat{T}_e \tag{2.1}$$

where $\hat{H}_{ion}$ is the ionic Hamiltonian and $\hat{T}_e$ denotes the kinetic energy operator of the Rydberg electron. From Bohr's model of hydrogen atom, the electron in a high Rydberg state moves slowly and its position is far away from the core. Therefore, as a zeroth-order description, we can neglect the kinetic energy term, and solve the ionic Schrödinger equation

$$\hat{H}_{ion}\Theta_a\left(\vec{R},\vec{r}_c;\vec{r}\right) = U_a\left(\vec{r}\right)\Theta_a\left(\vec{R},\vec{r}_c;\vec{r}\right) \tag{2.2}$$

where Θ_a and U_a represent the wavefunction and ion energy of the parent ion at state a, respectively. Here $\vec{R}$ and $\vec{r}_c$ denote the nuclear coordinate and the coordinates of the core electrons, respectively, while $\vec{r}$ denotes the coordinate of the Rydberg electron. Notice that both Θ_a and U_a depends on the Rydberg electron coordinate $\vec{r}$ parametrically. Using the completeness theorem, we can expand the total wavefunction Ψ in terms of Θ_a.

$$\Psi = \sum_a \Phi_a\left(\vec{r}\right)\Theta_a\left(\vec{R},\vec{r}_c;\vec{r}\right) \tag{2.3}$$

Substituting equation (2.3) into the full Schrödinger equation, we have

$$\sum_a \hat{T}_e\left(\Phi_a\Theta_a\right) + \sum_a U_a\left(\Phi_a\Theta_a\right) = E\sum_a \Phi_a\Theta_a \tag{2.4}$$

where

$$\begin{aligned}\hat{T}_e\left(\Phi_a\Theta_a\right) &= \Theta_a\hat{T}_e\Phi_a - \frac{\hbar^2}{2m_e}\left(2\nabla_e\Theta_a\cdot\nabla_e\Phi_a + \Phi_a\nabla_e^2\Theta_a\right)\\ &\equiv \Theta_a\hat{T}_e\Phi_a + \hat{H}'_{IBO}\left(\Phi_a\Theta_a\right)\end{aligned} \tag{2.5}$$

Multiplying Θ_b^* on both sides of equation (2.4), integrating over the ionic coordinates, and neglecting the terms involving $\hat{H}'_{IBO}$, we obtain

$$(\hat{T}_e + U_b)\Phi_{bm} = E_{bm}\Phi_{bm} \tag{2.6}$$

and

$$\Psi_{bm} = \Phi_{bm}\Theta_b \tag{2.7}$$

Here m denotes the quantum numbers to specify the Rydberg electronic state. We shall denote the ionic wavefunction in the core electronic state c and rovibrational state w as Θ_{cw} ; i.e., b=cw, and represent the corresponding Rydberg electronic wavefunction by Φ_{cwm}. From equations (2.2) and (2.6), we can see that the eigenvalue U_b of ionic Schrödinger equation serves as the potential energy of the electronic Schrödinger equation. It can also be interpreted that the Rydberg electron is moving on the potential energy surface provided by the ion core in the rovibronic state cw.

2.2 Perturbation and multipole expansion

In this section, we shall use the time independent perturbation theory to derive the zeroth-order wavefunction and higher order corrections. Consider a molecule containing n-1 electrons in the core and one electron in the high Rydberg state (denoted as the electron n). The ionic Hamiltonian can be expressed as

$$\hat{H}_{ion} = -\frac{\hbar^2}{2m_e}\sum_{i=1}^{n-1}\nabla_i^2 + \hat{T}_N + V \tag{2.8}$$

where $\hat{T}_N$ is the kinetic energy operator of the nuclei. Next, we can separate the interactions containing the electron n from V

$$\begin{aligned} V &= V_{ion} + \sum_{i=1}^{n-1}\frac{e^2}{r_{n,i}} - \sum_{\alpha}\frac{Z_\alpha e^2}{r_{n,\alpha}} \\ &\equiv V_{ion} + V' \end{aligned} \tag{2.9}$$

where i represents the i-th electron in the ion core, and α refers to α-th nucleus. V' can be approximated by the multipole expansion.

$$\begin{aligned} V' &= -\frac{e^2}{r_n} + \left[\sum_{i=1}^{n-1}\frac{e^2(\vec{r}_n\cdot\vec{r}_i)}{r_n^3} - \sum_{\alpha}\frac{Z_\alpha e^2(\vec{r}_n\cdot\vec{R}_\alpha)}{r_n^3}\right] + \left\{\sum_{i=1}^{n-1}\frac{e^2}{2}\left[\frac{3(\vec{r}_n\cdot\vec{r}_i)^2}{r_n^5} - \frac{r_i^2}{r_n^3}\right]\right. \\ &\quad \left. -\sum_{\alpha}\frac{Z_\alpha e^2}{2}\left[\frac{3(\vec{r}_n\cdot\vec{R}_\alpha)^2}{r_n^5} - \frac{R_\alpha^2}{r_n^3}\right]\right\} + \cdots \\ &\equiv -\frac{e^2}{r_n} + V_d' + V_q' + \cdots \end{aligned} \tag{2.10}$$

where V_d' denotes the dipole interaction and V_q' the quadrupole interaction, and so on. Neglecting the higher order interactions, we obtain the zeroth-order wavefunctions of the ion core and the Rydberg electron. That is, equations (2.2) and (2.6) can be expressed as

$$\hat{H}_{ion}^0\Theta_{cw}^0\left(\vec{R},\vec{r}_c\right) = U_{cw}^0\Theta_{cw}^0\left(\vec{R},\vec{r}_c\right) \tag{2.11}$$

$$(\hat{T}_e - \frac{e^2}{r_n})\Phi_m^0(\vec{r}_n) = \varepsilon_m^0 \Phi_m^0(\vec{r}_n) \tag{2.12}$$

Here we can see that the solution of equation (2.11) is the wavefunction of a bare ion, which can be calculated with the conventional quantum chemistry methods, while the solution of equation (2.12) is the wavefunction of a hydrogen atom. Therefore, as the conventional hydrogenic wavefunction [18], the quantum numbers *nlm* can be used to describe the bound electron, while *klm* can be used to describe the free electron; i.e., m=*nlm* or m=*klm*. Letting $\vec{r}_n = \vec{r} = (r, \theta, \phi)$ in spherical coordinates, we have

$$\Phi_{nlm}^0 = R_{nl}(r) Y_{lm}(\theta, \phi) \tag{2.13}$$

$$\Phi_{klm}^0 = R_{kl}(r) Y_{lm}(\theta, \phi) \tag{2.14}$$

To solve equation (2.11), we can use Born-Oppenheimer approximation and the ionic wavefunction Θ_{cw} can be separated as the core electronic wavefunction φ_c and the nuclear wavefunction ψ_{cw}. That is,

$$\hat{H}_{ion}^0 = \hat{H}_e^0 + \hat{T}_N \tag{2.15}$$

$$\hat{H}_e^0 \varphi_c^0 = u_c^0 \varphi_c^0 \tag{2.16}$$

$$\left(\hat{T}_N + u_c^0\right)\psi_{cw}^0 = U_{cw}^0 \psi_{cw}^0 \tag{2.17}$$

and

$$\Theta_{cw}^0 = \varphi_c^0 \psi_{cw}^0 \tag{2.18}$$

where $\hat{H}_e^0$ denotes the Hamiltonian of the core electrons after subtracting the nuclear kinetic energy operator $\hat{T}_N$ from $\hat{H}_{ion}^0$. In the case of diatomic molecules, the zeroth-order nuclear wavefunction ψ_{cw}^0 can be expressed in spherical coordinates $\vec{R} = (R, \Theta, \Phi)$ as

$$\psi_{cv^+N^+M_N^+}^0 = \chi_{cv^+}(R) Y_{N^+M_N^+}(\Theta, \Phi) \tag{2.19}$$

where χ_{cv^+} denotes the vibrational wavefunction and $Y_{N^+M_N^+}$ denotes the rotational wavefunction with respect to the electronic state c of the ion core. Here $w = v^+N^+M_N^+$ denote the vibration and rotation quantum numbers.

For a homonuclear diatomic molecule, the first-order energy correction of the dipole interaction is zero, so we shall perform the first-order correction of the quadrupole term. It follows that

$$u_c^{(1)} = \left\langle \varphi_c^0 \middle| V_q' \middle| \varphi_c^0 \right\rangle \tag{2.20}$$

Performing the integration with respect to the core electronic coordinates $\vec{r}_c$, we can obtain the conventional expression by using quadrupole moment Q as functions of the nuclear coordinate $\vec{R}$.

$$u_c^{(1)} = -e\frac{Q_c(R)}{r^3}\left[\frac{4\pi}{5}\sum_{m=-2}^{2} Y_{2m}(\theta,\phi)Y_{2m}^*(\Theta,\Phi)\right] \tag{2.21}$$

where the first spherical harmonics depends on the coordinate of the Rydberg electron, and the second one depends on that of the ion core.

2.3 The breakdown of IBOA and autoionization

The basic concept of autoionization can be understood by using the diagram in figure 1, which is based on the IBOA model. From equations (2.11) and (2.12), we can see that, in the high Rydberg state, the electron behaves like a hydrogenic electron, and the core like a bare ion. The interaction between them is very week, so we can form two independent systems, as shown in the zeroth-order Hamiltonian.

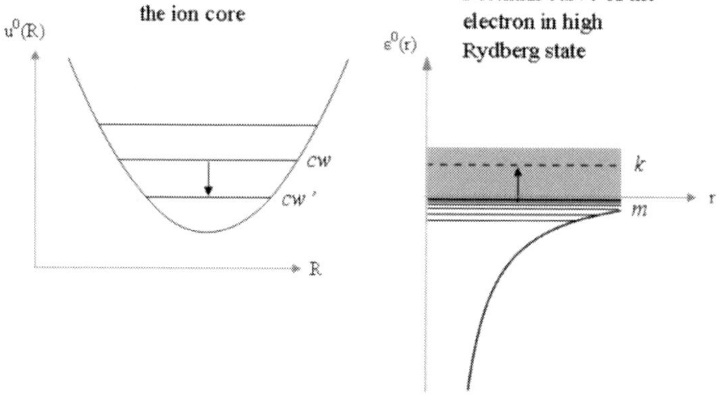

Figure 1. The diagram of vibrational autoionization based on the model of IBOA. During the autoionization $cwm \to cw'k$, the ion core would relax from a vibrationally excited state cw to a lower state cw', while the electron in the high Rydberg state m would autoionize to the continuum state k.

As shown in the figure, the molecule was first excited to a vibrationally or rotationally excited state with one electron in the high Rydberg state. Due to the interaction within the molecule, the excited core would release its energy and transfer the energy to the Rydberg electron yielding ionization. With this understanding, we can calculate the corresponding transition rate.

Lin [19] has pointed out that within the BOA the nuclear kinetic energy operator can be used to model the radiationless processes, such as internal conversion or intersystem crossing. Likewise, here we can use the kinetic energy operator of the Rydberg electron to study the radiationless process in high Rydberg states, such as autoionization [20]. To derive the expression of the transition rate, we need to include the neglected term in equation (2.5); in other words, the radiationless transition is interpreted as the breakdown of the IBOA. Therefore, to calculate the autoionization rate, we can use the Fermi's golden rule

$$W_{(cwm\to cw'k)} = \frac{2\pi}{\hbar}\left|\langle \Psi_{cw'k} | \hat{H}'_{IBO} | \Psi_{cwm} \rangle\right|^2 \rho(E_k) \tag{2.22}$$

where $\rho(E_k)$ is the density of states for the ionized electron of energy E_k and

$$\begin{aligned}&\left\langle \Psi_{cw'k} \middle| \hat{H}'_{IBO} \middle| \Psi_{cwm} \right\rangle \\ &= -\frac{\hbar^2}{2m_e}\left(2\left\langle \Phi_{cw'k} \middle| \left\langle \Theta_{cw'} \middle| \nabla_e \Theta_{cw} \right\rangle \cdot \middle| \nabla_e \Phi_{cwm} \right\rangle + \left\langle \Phi_{cw'k} \middle| \left\langle \Theta_{cw'} \middle| \nabla_e^2 \Theta_{cw} \right\rangle \middle| \Phi_{cwm} \right\rangle\right)\end{aligned} \tag{2.23}$$

To evaluate $\nabla_e \Theta_{cw}$ and $\nabla_e^2 \Theta_{cw}$, we shall consider the first order correction of the ionic wavefunction. Using (2.19) and (2.21), we have

$$\begin{aligned}\left\langle \Theta_{cv^{+\prime}N^{+\prime}M_N^{+\prime}} \middle| \nabla_e \middle| \Theta_{cv^+N^+M_N^+} \right\rangle &= \frac{\left\langle \Theta^0_{cv^{+\prime}N^{+\prime}M_N^{+\prime}} \middle| \nabla_e V_q' \middle| \Theta^0_{cv^+N^+M_N^+} \right\rangle}{U^0_{cv^+N^+} - U^0_{cv^{+\prime}N^{+\prime}}} \\ &= -e\frac{\left\langle \chi_{cv^{+\prime}} \middle| Q_c \middle| \chi_{cv^+} \right\rangle}{U^0_{cv^+N^+} - U^0_{cv^{+\prime}N^{+\prime}}}\left\{\frac{4\pi}{5}\sum_{\bar{m}=-2}^{2}\left\langle Y_{N^{+\prime}M_N^{+\prime}} \middle| Y^*_{2\bar{m}} \middle| Y_{N^+M_N^+} \right\rangle \nabla_e\left[\frac{Y_{2\bar{m}}(\theta,\phi)}{r^3}\right]\right\}\end{aligned} \tag{2.24}$$

$$\left\langle \Theta_{cv^{+\prime}N^{+\prime}M_N^{+\prime}} \middle| \nabla_e^2 \middle| \Theta_{cv^+N^+M_N^+} \right\rangle = \frac{\left\langle \Theta^0_{cv^{+\prime}N^{+\prime}M_N^{+\prime}} \middle| \nabla_e^2 V_q' \middle| \Theta^0_{cv^+N^+M_N^+} \right\rangle}{U^0_{cv^+N^+} - U^0_{cv^{+\prime}N^{+\prime}}} = 0 \tag{2.25}$$

Here $w' = v^{+\prime}N^{+\prime}M_N^{+\prime}$, $k = kl'm'$, $w = v^+N^+M_N^+$ and $m = nlm$, respectively. The electronic part of the quadrupole interaction V_q' is a solution of the Laplace equation, and thus the matrix element in equation (2.25) will vanish. The integration over the Rydberg electronic coordinate can be performed using the recursion relations of the hydrogenic wavefunction. For spherical harmonics, we obtain

$$\left\langle Y_{l'm'} \middle| \frac{\partial Y_{2\bar{m}}}{\partial\theta} \middle| \frac{\partial Y_{lm}}{\partial\theta} \right\rangle + \left\langle Y_{l'm'} \middle| \frac{1}{\sin\theta}\frac{\partial Y_{2\bar{m}}}{\partial\phi} \middle| \frac{1}{\sin\theta}\frac{\partial Y_{lm}}{\partial\phi} \right\rangle = D_{l',l}\left\langle Y_{l'm'} \middle| Y_{2\bar{m}} \middle| Y_{lm} \right\rangle \tag{2.26}$$

where $D_{l,l} = 3$, $D_{l+2,l} = -2l$, and $D_{l-2,l} = 2(l+1)$.

Therefore, we obtain the final expression of the transition matrix element as

$$\begin{aligned}&\left\langle kl' v^{+\prime} N^{+\prime} J\, M_J \middle| \hat{H}'_{IBOA} \middle| nl\, v^+ N^+ J\, M_J \right\rangle \\ &= -\frac{\hbar^2}{m_e}\frac{(-e)}{U^0_{cv^+N^+} - U^0_{cv^{+\prime}N^{+\prime}}}\left\langle \chi_{cv^{+\prime}} \middle| Q_c(R) \middle| \chi_{cv^+} \right\rangle \times \left[D_{l',l}\left\langle R_{kl'} \middle| r^{-5} \middle| R_{nl} \right\rangle - 3\left\langle R_{kl'} \middle| r^{-4} \middle| \frac{dR_{nl}}{dr} \right\rangle\right] \\ &\times g\left(l', N^{+\prime}, J', M_J'; l, N^+, J, M_J\right)\end{aligned} \tag{2.27}$$

where *nl* and *kl'* represent the quantum numbers of discrete states and that of continuum states of the ZEKE electron, respectively. The angular matrix element *g* is shown as in [21, 22]

$$g\left(l',N^{+'},J',M_J';l,N^{+},J,M_J\right)=\left(-1\right)^{l+l'+J}\begin{Bmatrix} J & N^{+'} & l' \\ 2 & l & N^{+} \end{Bmatrix}\delta_{J',J}\delta_{M_J',M_J}$$
$$\times\left[\left(2l'+1\right)\left(2l+1\right)\right]^{1/2}\begin{pmatrix} l' & 2 & l \\ 0 & 0 & 0 \end{pmatrix}$$
$$\times\left[\left(2N^{+'}+1\right)\left(2N^{+}+1\right)\right]^{1/2}\begin{pmatrix} N^{+'} & 2 & N^{+} \\ 0 & 0 & 0 \end{pmatrix} \tag{2.28}$$

In equation (2.27), there are three matrix elements to be calculated. The first matrix element contains the quadrupole moment of the ion core, which is a function of internuclear distance R. It is responsible for the vibrational transition. As the propensity rule of vibrational autoionization, the magnitude of the corresponding transition rate would decrease rapidly as Δv increases [23, 24]. The second one is the radial integral of the Rydberg electron, and it describes the transition from an electronic bound state to the continuum. In this section, we employed the recursion relation of associated Laguerre polynomial and performed the numerical integration to obtain the value of this matrix element. The last one involves the angular momentum for both electronic and nuclear rotation motions. Due to the symmetry of spherical harmonics, the allowed transitions should obey $\Delta l = 0,\pm 2$ and $\Delta N^{+} = 0,\pm 2$. In addition, the projection quantum numbers m and M_N^{+} are summed for the coupling between the electronic orbital angular momentum l and the core rotational angular momentum N^{+}. Then, we have the total angular momentum J and its projection quantum number M_J. Due to the Kronecker delta, J and M_J are constant upon the transition. The selection rules and propensity rule of this formulation are in agreement with the previous studies of autoionization based on the multipole interaction [22] and the MQDT [25, 26].

3. Result and discussion

Here we shall calculate the autoionization rate of H_2, with respect to the transition among different rotational and vibrational states with transition $\left\{nl\ ;\ v^{+}N^{+};\ J\ M_J\right\}\rightarrow\left\{kl'\ ;\ v^{+'}N^{+'};\ J\ M_J\right\}$.

As shown in figure 1, there is an energy restriction for autoionization, which is due to the Dirac delta function in the Fermi's golden rule. Nevertheless, in order to analyze the relation between the magnitude of transition rate and the corresponding quantum numbers, we first neglect this restriction. In the following, we will consider the transitions for different n, l and energy of the unbound state; the related transition for vibrational and rotational states is $\left(v^{+}=1,\ N^{+}=0\right)\rightarrow\left(v^{+'}=0,\ N^{+'}=2\right)$.

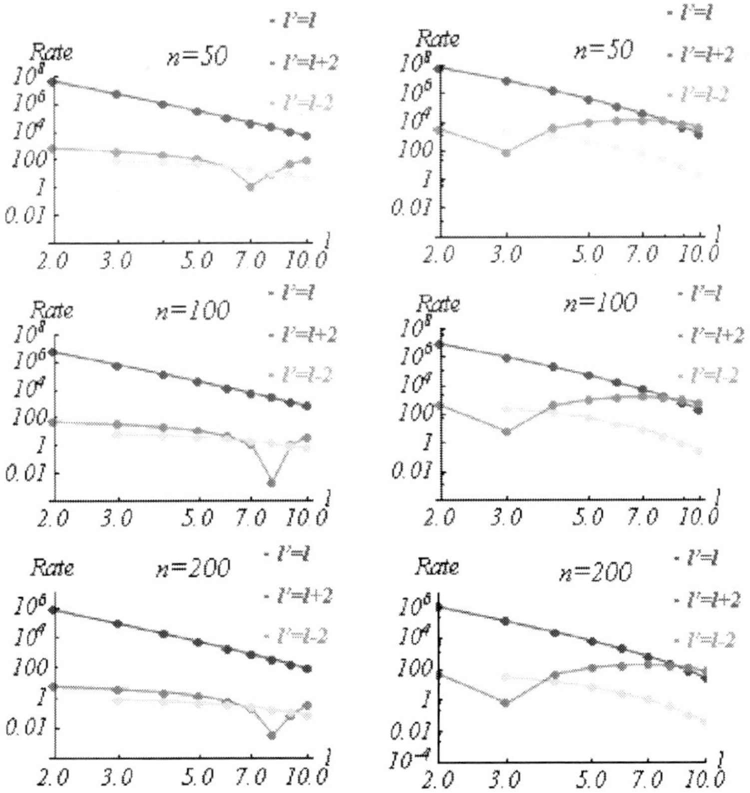

Figure 2. Comparison of the calculated autoionization rate for $l'=l$, $l'=l+2$, and $l'=l-2$ in terms of the quadrupole interaction. The left panel is for the continuum-state energy of 100 cm^{-1}; the right panel is for the energy of 1000 cm^{-1}.

The numerical results calculated using equation (2.27) for H_2 are given in figure 2. If we specify the quantum numbers n and k, there are three possibilities of the final l' which obey the selection rule. First, in the case of $l'=l$, we can see that the rate decreases rapidly with increasing l. For example, the corresponding rates of n=100 can range from 10^3 to 10^7 s^{-1}. This behavior can be understood by the property of the hydrogenic wavefunction [18]. That is, the states at low l would have larger core penetration as well as stronger interaction, and thus the rate would decrease as l increases. Likewise, the related rate would decrease as n increases due to the weak interaction between the Rydberg electron and ion core. As for the k dependence, we found that the transition rates for low-l states are relatively invariant to the influence of k. However, those of high-l states have stronger energy dependence as the energy increases.

Compared with the transitions of $l'=l$, those of $l'=l+2$ and $l'=l-2$ are relatively slow. However, for the energy of 1000 cm^{-1}, the transitions of $l'=l+2$ would compete with those of $l'=l$ at larger l (l>8). Besides, from figure 2, we can see that there is no obvious tendency for these two cases. For $l'=l-2$, the rate would decrease with increasing l quite slowly. On the other hand, for $l'=l+2$, we can find a dip for a specific l in each plot. For both cases, the magnitudes of the autoionization rates are more sensitive to the energy than $l'=l$.

From figure 2, it is clear that the transition of $l'=l$ shows a regular power law dependence of l. Using a nonlinear modeling, we obtain the scaling law for the autoionization rate as

$$rate = 3.648 \times 10^{14} n^{-3} l^{-5.3}\ s^{-1} \tag{3.1}$$

The n^3 law is consistent with that obtained from the asymptotic wavefunction of high Rydberg state. The comparison between the calculated data and the fitting curves is shown in figure 3.

In figure 3, we can see the numerical results agree well with the estimation of the scaling law for the low energy case. In equation (3.1), the scaling law is independent of the energy of the continuum state, but it still works well for the low-l transitions when the energy is relatively high. For instance, it can effectively predict the rate up to l=5 for the energy of 1000 cm^{-1}. On the other hand, for the transitions $l' = l \pm 2$, we cannot derive a scaling law of the transition rate due to the irregular dependence on l. Nevertheless, for most cases, l'=l is the most probable transition, so we can neglect the contribution of $l' = l \pm 2$ transitions. In this sense, this scaling law is quite useful to gain an approximate magnitude of the autoionization rate.

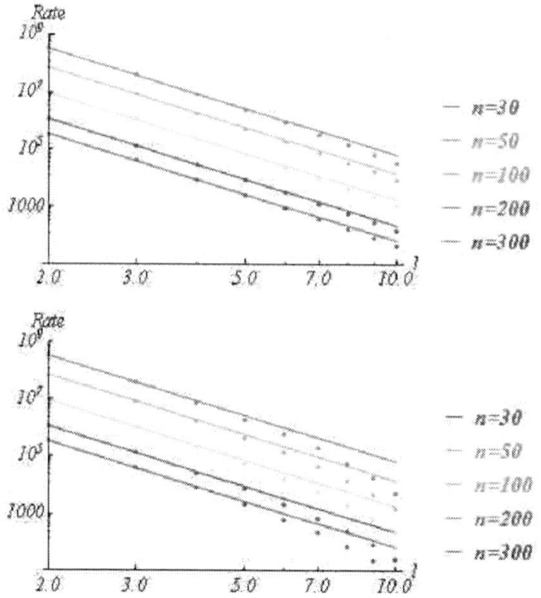

Figure 3. Comparison of the autoionization rate for Rydberg state (n,l) based on the quadrupole interaction. The upper panel is for the continuum-state energy of 100 cm^{-1}; the lower panel is for the energy of 1000 cm^{-1}. The calculated rate based on the breakdown of the IBOA is marked by point for each n, l; the lines represent the rates obtained from the scaling law.

Acknowledgements
This work is partly supported by the NSC of ROC and CQSE of NTU 97R0066-66.

References

[1] Müller-Dethlefs K, Sander M and Schlag E W 1984 *Chem. Phys. Lett.* **112** 291
[2] Schlag E W 1998 *ZEKE Spectroscopy* (Cambridge: Cambridge University Press)
[3] Cockett M C R 2005 *Chem. Soc. Rev.* **34** 935
[4] Willitsch S, Wuest A and Merkt F 2004 *Chimia* **58** 281
[5] Müller-Dethlefs K and Schlag E W 1991 *Ann. Rev. Phys. Chem.* **42** 109

[6] Merkt F 1997 *Annu. Rev. Phys. Chem.* **48** 673
[7] Ng C-Y 2002 *Annu. Rev. Phys. Chem.* **53** 101
[8] Wright T 2002 *Annu. Rep. Prog. Chem. Sect. C* **98** 375
[9] Neusser H J and Krause H 1994 *Chem. Rev.* **94** 1829
[10] Dessent C E H and Müller-Dethlefs K 2000 *Chem. Rev.* **100** 3999
[11] Merkt F and Softley T P 1993 *Int. Rev. Phys. Chem.* **12** 205-239
[12] Softley T P 2004 *Int. Rev. Phys. Chem.* **23** 1-78
[13] Chao S D, Hayashi M, Lin S H and Schlag E W 1998 *J. Phys. B* **32** 2007
[14] Chao S D, Hayashi M, Lin S H and Schlag E W 1998 *J. Chin. Chem. Soc.* **45** 491
[15] Even U, Ben-Nun M and Levine R D 1993 *Chem. Phys. Lett.* **210** 416
[16] Remacle F and Levine R D 1998 *Int. J. Quan. Chem.* **67** 85
[17] Chao S D 1999 *Thesis* (Taipei: National Taiwan University)
[18] Bethe H A and Salpeter E E 1957 *Quantum Mechanics of One- and Two- Electron Atom* (New York: Academic Press Inc.)
[19] Lin S H 1966 *J. Chem. Phys.* **44** 3759
[20] Chao S D, Selzle H L, Neusser H J, Schlag E W, Yao L and Lin S H 2007 *Z. Phys. Chem.* **221** 633-646
[21] Bixon M and Jortner J 1996 *Mol. Phys.* **89** 373-401
[22] Eyler E E and Pipkin F M 1983 *Phys. Rev.* A **27** 2462
[23] Bates D R and Poots G 1953 *Proc. Phys. Soc. (London)* **A66** 784
[24] Berry R S 1966 *J. Chem. Phys.* **45** 1128
[25] Herzberg G and Jungen Ch 1972 *J. Mol. Spectrosc.* **41** 425
[26] Raoult M and Jungen Ch 1981 *J. Chem. Phys.* **74** 3388

International Workshop on Electronic Spectroscopy for Gas-phase Molecules and Solid Surfaces IOP Publishing
Journal of Physics: Conference Series **235** (2010) 012003 doi:10.1088/1742-6596/235/1/012003

Scanning Tunneling Microscopy and Inelastic Electron Tunneling Spectroscopy Studies of Methyl Isocyanide Adsorbed on Pt(111)

Satoshi Katano[1,2,*], Yousoo Kim[2], Masafumi Hori[2], Michael Trenary[3] and Maki Kawai[2,4,*]

[1] Research Institute of Electrical Communication, Tohoku University, 2-1-1 Katahira, Aoba-ku, Sendai 980-8577, Japan
[2] Surface Chemistry Laboratory, RIKEN, 2-1 Hirosawa, Wako, Saitama 351-0198 Japan
[3] Department of Chemistry, University of Illinois at Chicago, 845 West Taylor Street, Chicago, Illinois 60607-7061 USA
[4] Department of Advanced Materials Science, University of Tokyo, 5-1-5 Kashiwanoha, Kashiwa, Chiba 277-8651, Japan

Abstract. A low-temperature scanning tunneling microscope (STM) was used to investigate the adsorption state of a single methyl isocyanide (MeNC) molecule on the Pt(111) surface at 4.7 K. We found that MeNC was resolved as a round-shaped protrusion in the STM image. The STM image of paired MeNC is highly protruded in comparison with that of isolated MeNC due to the charge transfer from Pt to MeNC. Inelastic electron tunneling spectroscopy with the STM system (STM-IETS) was also employed in order to reveal the adsorption state of individual MeNC molecules on Pt(111). The STM-IETS spectrum of MeNC exhibits peaks at 8, 48 and 375 mV. Referring to the vibrational spectra reported previously, we assigned these peaks to the frustrated translation mode, PtC stretching mode and CH_3 stretching mode, respectively. The absence of other vibrational modes could be due to a reduction of the elastic tunneling current.

1. Introduction

Scanning tunneling microscopy (STM) opens a novel opportunity to explore individual molecules at the atomic scale. Recently, inducing motion and reaction of single molecules adsorbed on metal surfaces has been extensively studied using STM in connection with the ultimate goal of developing nanometer-scale molecular electronics.[1,2] Various types of excitation mechanism, including vibrational[3-9] and electronic excitations[10-18] and mechanical interaction[19-21] have been investigated for controlling the dynamical behavior of a single adsorbed molecule.

On the other hand, STM is also capable of spectroscopic measurements at high resolution. Particularly, inelastic electron tunnelling spectroscopy based on STM (STM-IETS) provides a vibrational spectrum of a single molecule.[1,2,6] We can directly analyze the chemical properties of individual atoms and molecules, which are the key to the practical behavior of materials.

* To whom any correspondence should be addressed: maki@riken.jp (M. K.), skatano@riec.tohoku.ac.jp (S. K.)

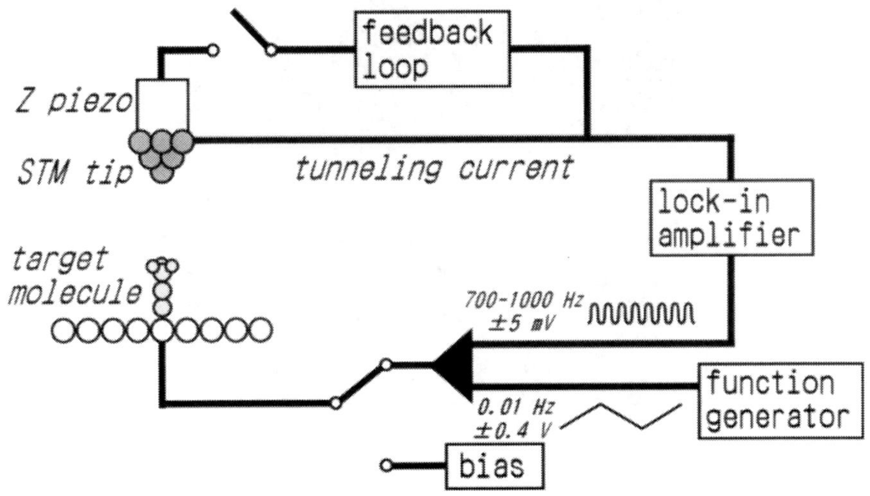

Figure 1. Schematic diagram of IETS measurements with the STM.

In this study, the observation of single molecules of methyl isocyanide (MeNC) on Pt(111) was carried out using low-temperature STM. MeNC and other isocyanide derivatives have been studied on transition metal surfaces because isocyanide (C≡N-) is isoelectronic with carbon monoxide (CO), which is the most studied adsorbate in the field of surface science.[22] Furthermore, isocyanide derivatives have attracted considerable attention in the context of molecular electronics since isocyanide offers a favorable electrical contact through the NC-metal bond. [23-26] We recently achieved the reversible reaction of MeNC on Pt(111) using the STM tip.[17] MeNC reacts with a hydrogen atom to form MeHNC through the exposure of H_2 gas. Subsequently, MeHNC is converted back to the original MeNC by the application of a voltage pulse from the STM tip. These reactions that drastically alter the electronic hybridization between metal and molecule are of great importance for the realization of novel functionality in molecular electronics. Here, we found that pairs of MeNC molecules appear as higher protrutions in the STM image compared to isolated MeNC molecules. The observed difference in height is associated with the degree of charge transfer from Pt to MeNC. STM-IETS was also employed in order to reveal the adsorption state of individual MeNC molecules on Pt(111). Vibrational peaks appearing in the STM-IETS spectrum are compared with the peaks observed in the previously reported vibrational spectra.

2. Experimental

All experiments were performed in an ultrahigh vacuum (UHV) chamber with a base pressure of 4×10^{-11} Torr. The Pt(111) surface was cleaned by repeated Ar ion sputtering and annealing at 1000 K, until the clean surface was confirmed by STM imaging. Methyl isocyanide (CH_3NC; MeNC) was prepared by the method described by Casanova et al.[27] The purity of the synthesized MeNC was over 99 % as checked by a gas chromatograph mass spectrometer. The clean Pt(111) surface was exposed to MeNC at about 50 K through a dosing tube located near the substrate. STM experiments were carried out using a low-temperature STM system (LT-STM, Omicron GmbH). The STM images shown here were obtained at 4.7 K in constant-current mode. A schematic diagram of the IETS measurement is shown in Figure 1. The STM tip was precisely positioned above the target molecule followed by turning off the feedback loop. Then, the bias voltage was swept (0.01 Hz, ±0.4 V) by the external system using a function generator. The second derivative of the tunneling current (d^2I/dV^2) was sensitively detected by a lock-in amplifier using a modulation frequency of 797 Hz and a modulation voltage of 5 mV.

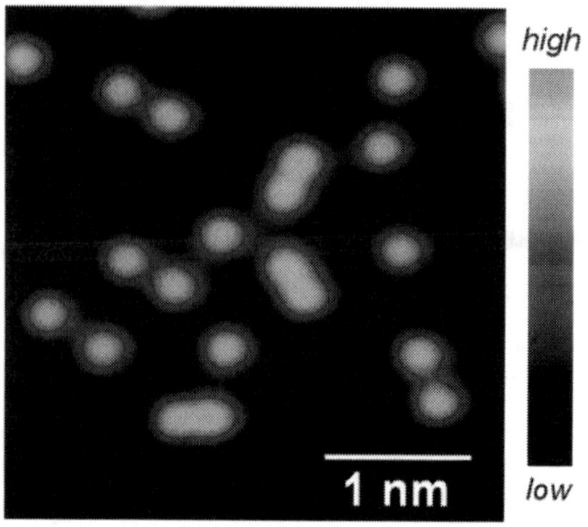

Figure 2. STM image of MeNC adsorbed on Pt(111) at 4.7 K. The image was obtained with a sample bias of 0.02 V and a tunneling current of 1.0 nA.

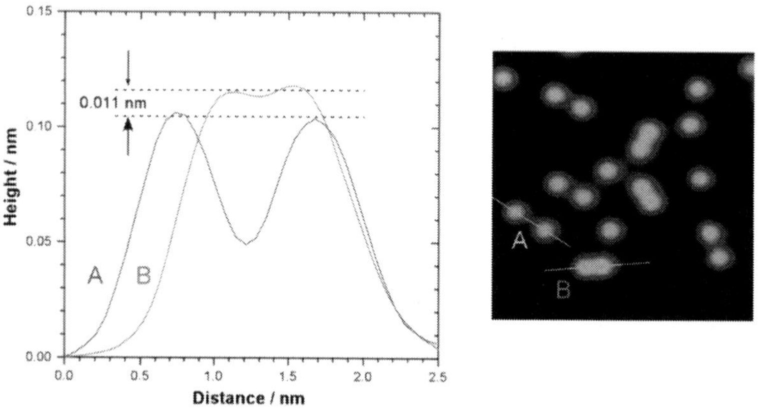

Figure 3. Height profiles of (A) isolated and (B) paired MeNC molecules adsorbed on Pt(111). Corresponding STM image of the isolated monomer (A) and dimer (B) is shown on the right.

3. Results and discussion

Figure 2 shows an STM image of Pt(111) at 4.7 K obtained after exposing to MeNC at 50 K. Single MeNC molecules are resolved as protrusions. Other STM images, showing the atomic structure of the Pt substrate, revealed that MeNC is adsorbed at the on-top sites [28], which was also suggested from studies using conventional vibrational spectroscopies.[22,29] Another type of bonding configuration appears when increasing the surface coverage of MeNC. We found that MeNC is adsorbed at the bridge site as well as the on-top site at high coverages.[28] Such a transformation of the bonding site is similarly observed for the adsorption of CO, which is isoelectronic with MeNC.[30,31]

As seen from Figure 2, MeNC was dispersed on the whole Pt surface when dosed at 50 K. Careful STM image analysis indicates that MeNC molecules are likely to form pairs. Consecutive STM imaging demonstrated that MeNC preferentially forms pairs of molecules with distances of $\sqrt{3}a$ (0.48 nm) and $\sqrt{7}a$ (0.73 nm), where a represents the closest Pt-Pt distance (0.2775 nm) of the Pt(111) surface.[28] We found that paired MeNC molecules give a higher protrusion in the STM image than isolated MeNC molecules. The height profiles of MeNC on Pt(111) revealed that the MeNC pairs

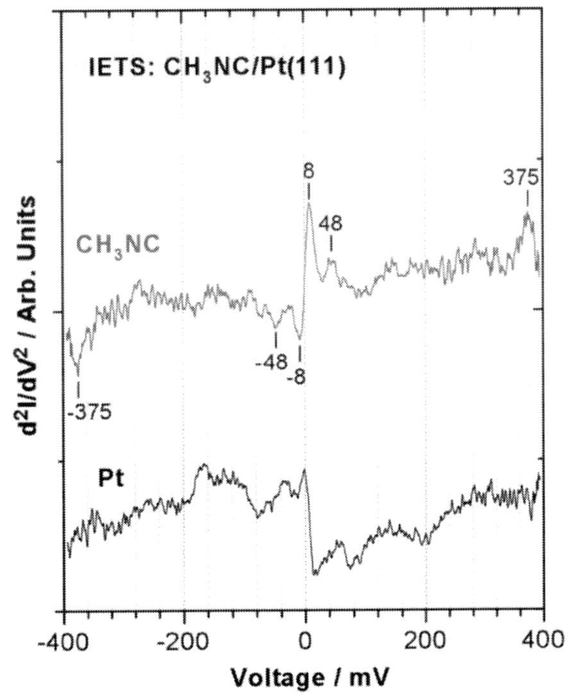

Figure 4. STM-IETS spectrum of MeNC adsorbed on Pt(111) at 4.7 K. The background spectrum measured at the Pt surface is also shown in the bottom. All spectra were obtained after the feedback loop was turned off at the voltage of 0.1 V and the tunneling current of 0.4 nA.

(separation distance of $\sqrt{3}a$) are 0.011 nm higher than the isolated MeNC molecules, as shown in Figure 3. The observed height difference can not be explained only from the convolution of the height of the neighbour MeNC molecules. It should be noted that the formation of MeHNC from the hydrogenation of MeNC is also accompanied by a change in the apparent height. However, the height of MeHNC observed in the STM image is 0.021 nm lower than a single isolated MeNC molecule.[17]

The observed difference in molecular height between paired and isolated MeNC is probably caused by the charge transfer between MeNC and Pt. Avery *et al.* reported that the adsorption of MeNC on metal surfaces is described by σ donation (electron transfer from the 5σ orbital of MeNC to the empty d orbital of the metal) and π back-donation (electron transfer from the occupied d orbital of the metal to the $2\pi^*$ orbital of MeNC),[22] which is similar to the case of CO adsorbed on metal surfaces. Such a charge transfer between molecule and metal is known to affect the molecule's vibrational energy since the occupation of electrons in the antibonding electronic state of MeNC ($2\pi^*$) leads to the weakening of the CN bond strength. Kang *et al.* reported the vibrational spectra of MeNC on Pt(111) at various MeNC surface coverages.[29] They showed that the vibrational energy of the CN stretching mode is shifted to lower energy with increasing MeNC coverage. This indicates that charge transfer from Pt to MeNC increases as the distance between neighboring MeNC molecules decreases. Therefore, we conclude that the variation of the density of states near the Fermi level influences the apparent height in the STM image.

We carried out single molecule vibrational spectroscopy of MeNC using STM-IETS. The changes in tunneling conductance measured by STM-IETS can provide a direct measurement of the fraction of inelastically tunnelled electrons that excite the vibrational modes.[6] Figure 4 shows STM-IETS spectra of MeNC/Pt(111) and of clean Pt(111) at 4.7 K. Each STM-IETS spectrum was

Table 1. Assignments of the STM-IETS peaks from MeNC adsorption on Pt(111)[a]

Mode	Gas phase[b]	HREELS[c]	RAIRS[d]	IETS
ν(CH_3)	377 (a) 371 (s)	379	366	375
ν(CN)	271	280-283	280	—
δ(CH_3)	183 (a) 179 (s)	181	177	—
ν(PtC)	—	48	—	48
Frustrated translation	—	—	—	8

[a] The vibration energies are expressed in meV (1 meV = 8.06548cm^{-1}).
[b] Reference 32, [c] Reference 22, [d] Reference 29.

obtained after fixing the STM tip above the center of the molecule or above the substrate with the set voltage of 0.1 V and a set tunneling current of 0.4 nA followed by turning off the feedback loop. The STM-IETS spectrum of the Pt substrate exhibits a fine structure, which usually comes from the electronic structures of the STM tip. We would like to focus on the differences in characteristics between the IETS spectra obtained above the molecule and above the metal substrate. The STM-IETS spectrum of MeNC exhibits peaks at 8 and 48 mV for both positive and negative bias voltages near the Fermi level. These peaks are absent in the Pt spectrum at the corresponding energies. Therefore, the peaks observed at 8 and 48 mV are readily assigned to the frustrated translational (FT) mode and PtC stretching mode (ν(PtC)) of MeNC/Pt(111).[17]

In addition, a sharp peak appeared at 375 mV in the STM-IETS spectrum of MeNC. This peak was also observed in the negative bias region, indicating an inelastic tunneling electron-induced excitation of a vibrational mode. Referring to the vibrational assignment reported previously, the peak at 375 mV corresponds to the CH_3 stretching mode (ν(CH_3)).[22,29] The assignments of the observed peaks in the STM-IETS spectrum are summarized in Table 1 along with the assignments of vibrational peaks observed for gas phase MeNC and adsorbed MeNC as measured by other vibrational spectroscopy techniques.[22,29,32]

It is interesting to note that several peaks, i.e., CN stretching (ν(CN)) and CH_3 deformation (δ(CH_3)), were absent in the STM-IETS spectrum. In particular, the ν(CN) mode gives an intense peak in the HREELS and RAIRS spectra. Although the ν(CN) peak is expected to appear at 280-283 mV, the STM-IETS spectrum of MeNC has no feature in this energy region. Since MeNC is adsorbed on Pt(111) by σ donation and π back-donation, the vibrational mode of ν(CN) is strongly coupled with the electronic states near the Fermi level. Previously, Lorente *et al.* investigated the inelastic tunneling process via the vibrational excitation at the STM junction of acetylene on Cu(100) using a density functional theory (DFT) calculation.[33,34] Total conductance changes, which are observable in STM-IETS, have both inelastic and elastic tunneling contributions. Note that an increase in the conductance via the inelastic process is occasionally accompanied by a reduction in elastic tunneling. Thus, the FT, ν(PtC) and ν(CH_3) modes produced a gain in the inelastic tunneling current that gave a dominant IET signal. However, the total conductance change of the ν(CN) and δ(CH_3) modes must be negligible due to a reduction of elastic tunneling current.

4. Conclusion

We have investigated the adsorption of single methyl isocyanide (MeNC) molecules on Pt(111) using low temperature STM at 4.7 K. MeNC was resolved as a round-shaped protrusion in the STM image.

We found that paired MeNC molecules gave a higher protrusion than an isolated MeNC molecule. This is probably due to the fact that the charge transfer from Pt to MeNC is dominant for the paired MeNC. The STM-IETS spectrum of MeNC exhibits peaks at 8, 48 and 375 mV, which were assigned to the vibrational excitation of the FT, ν(PtC) and $\nu(CH_3)$ modes respectively. The absence of other peaks can be ascribed to the fact that the total conductance change of the ν(CN) and $\delta(CH_3)$ modes must be negligible due to reduction of the elastic tunneling current.

Acknowledgments
The present work was supported in part by the Grant-in-Aid for Scientific Research on Young Scientists B (No. 16750022), in part by the Grant-in-Aid for Scientific Research on Priority Areas, "Electron transport through a linked molecule in nano-scale" from the Ministry of Education, Culture, Sports, Science and Technology, Japan. M. T. acknowledges support from the United States National Science Foundation under grant CHE-0714562.

References
[1] Ho W 2002 *J. Chem. Phys.* **117** 11033.
[2] Komeda T 2005 *Prog. Surf. Sci.* **78** 41.
[3] Stipe B C, Rezaei M A and Ho W 1998 *Science* **279** 1907.
[4] Stipe B C, Rezaei M A, Ho W, Gao S, Persson M and Lundqvist B I 1997 *Phys. Rev. Lett.* **78** 4410.
[5] Stipe B C, Rezaei M A and Ho W 1998 *Phys. Rev. Lett.* **81** 1263.
[6] Stipe B C, Rezaei M A and Ho W 1998 *Science* **280** 1732.
[7] Kim Y, Komeda T and Kawai M 2002 *Phys. Rev. Lett.* **89** 4.
[8] Sainoo Y, Kim Y, Okawa T, Komeda T, Shigekawa H and Kawai M 2005 *Phys. Rev. Lett.* **95** 4.
[9] Kumagai T, Kaizu M, Hatta S, Okuyama H, Aruga T, Hamada I and Morikawa Y 2008 *Phys. Rev. Lett.* **100** 166101.
[10] Bartels L, Meyer G, Rieder K H, Velic D, Knoesel E, Hotzel A, Wolf M and Ertl G 1998 *Phys. Rev. Lett.* **80** 2004.
[11] Gaudioso J, Lee H J and Ho W 1999 *J. Am. Chem. Soc.* **121** 8479.
[12] Lauhon L J and Ho W 2000 *J. Phys. Chem. A* **104** 2463.
[13] Lauhon L J and Ho W 2000 *Phys. Rev. Lett.* **84** 1527.
[14] Hla S W, Bartels L, Meyer G and Rieder K H 2000 *Phys. Rev. Lett.* **85** 2777.
[15] Komeda T, Kim Y, Fujita Y, Sainoo Y and Kawai M 2004 *J. Chem. Phys.* **120** 5347.
[16] Lastapis M, Martin M, Riedel D, Hellner L, Comtet G and Dujardin G 2005 *Science* **308** 1000.
[17] Katano S, Kim Y, Hori M, Trenary M and Kawai M 2007 *Science* **316** 1883.
[18] Katano S, Kim Y, Kitagawa T and Kawai M 2008 *Jpn. J. Appl. Phys.* **47** 6156.
[19] Eigler D M and Schweizer E K 1990 *Nature* **344** 524.
[20] Bartels L, Meyer G and Rieder K H 1997 *Phys. Rev. Lett.* **79** 697.
[21] Bartels L, Meyer G and Rieder K H 1997 *Chem. Phys. Lett.* **273** 371.
[22] Avery N R and Matheson T W 1984 *Surf. Sci.* **143** 110.
[23] Chen J, Calvet L C, Reed M A, Carr D W, Grubisha D S and Bennett D W 1999 *Chem. Phys. Lett.* **313** 741.
[24] Seminario J M, De la Cruz C E and Derosa P A 2001 *J. Am. Chem. Soc.* **123** 5616.
[25] Kiguchi M, Miura S, Hara K, Sawamura M and Murakoshi K 2006 *Appl. Phys. Lett.* **89** 213104.
[26] Angelici R J and Lazar M 2008 *Inorg. Chem.* **47** 9155.
[27] Casanova J, Schuster R E and Werner N D 1963 *Journal of the Chemical Society* 4280.
[28] Katano S, Herceg E, Trenary M, Kim Y and Kawai M 2006 *J. Phys. Chem. B* **110** 20344.
[29] Kang D H and Trenary M 2002 *J. Phys. Chem. B* **106** 5710.
[30] Crossley A and King D A 1977 *Surf. Sci.* **68** 528.
[31] Avery N R 1982 *Appl. Surf. Sci.* **13** 171.
[32] Shimanouchi T: *Tables of Molecular Vibrational Frequencies*, NSRDS-NBS, U. S., 1972.

Journal of Physics: Conference Series **235** (2010) 012003 doi:10.1088/1742-6596/235/1/012003

[33] Lorente N and Persson M 2000 *Phys. Rev. Lett.* **85** 2997.
[34] Lorente N, Persson M, Lauhon L J and Ho W 2001 *Phys. Rev. Lett.* **86** 2593.

Journal of Physics: Conference Series **235** (2010) 012004 doi:10.1088/1742-6596/235/1/012004

Dynamics of single or double ionization of small systems from coincidence electron impact experiments

A Lahmam-Bennani[1], E M Staicu Casagrande and A Naja
Laboratoire des Collisions Atomiques et Moléculaires (LCAM, UMR8625),
Université Paris-Sud 11, F-91405 Orsay cedex, France

E-mail: azzedine.bennani@u-psud.fr

Abstract. In this brief review, we illustrate the potentialities and the power of electron-electron coincidence studies to investigate the dynamics of single ionization SI [(e,2e) case] or double ionization DI [(e,3e) case] processes. An example for (e,2e) SI of rare gas atoms is presented with a new insight into the behaviour of the recoil versus the binary intensity. A second example for (e,2e) SI of molecules is used to illustrate the observation of a purely molecular effect, namely the signature of interference effects due to the two-center nature of the H_2 molecule. The third example discusses an unprecedented triple coincidence study involving the scattered – ejected – Auger electrons emitted from an argon target. It is shown that the method allows to disentangle the contribution of various DI mechanisms.

1. Introduction

In an electron impact single ionization (SI) or double ionization (DI) experiment, an incident electron (indexed 0) scatters off an atomic or molecular target provoking the emission of one or two electrons from the target. Though indistinguishable, these electrons are indexed for convenience "a" for the fast-scattered one and "b" and "c" for the slower-ejected ones. A kinematically complete experiment is realized when these electrons are energy and angle selected and are detected in coincidence to insure that they are issued from the same ionizing event. These are the so-called (e,2e) [1] or (e,3e) [2] experiments, relative to the SI or DI cases, respectively. An impressive amount of work has been published to-date concerning the (e,2e) experiments, see for instance the reviews [3-5] and to a lesser extent concerning the (e,3e) experiments, see for instance the reviews [3,6]. In these studies one might grossly distinguish three kinematical regimes: (i) The dipolar regime, reached at high impact energy (several keV) and very small momentum transfer to the target (few tenths of an au); here the incident projectile acts as a pseudo photon. (ii) The Bethe ridge regime (also called electron momentum spectroscopy (EMS) regime), reached when the momentum transferred to the target is fully absorbed by the ejected electron, the nucleus acting as a spectator. Here the experiments are used to gather information about the target structure. And (iii) the more general case where neither of the above two conditions is fulfilled. The experiments are then used to investigate the dynamics of the SI or the DI process. In this paper, we will briefly discuss a few examples which all belong to this third category, i.e. the kinematics is far away from both the dipolar limit and from the Bethe ridge. These examples are meant to illustrate the very wide possibilities offered by both (e,2e) and (e,3e) techniques to investigate in the finest details some dynamical aspects of electron impact ionization studies.

---

[1] To whom any correspondence should be addressed.

The experimental set up used for these investigations has been fully described elsewhere [7]. It essentially features multi-angle analysis and double/triple coincidence detection of the two/three final state electrons, based on three toroidal analysers coupled with three position sensitive detectors (PSD). The key point is that the angular distribution of the scattered or ejected electrons in the collision plane is preserved upon arrival on the PSD's.

2. (e,2e) results on simple atoms

We first discuss results for SI of the rare gas atoms He, Ne, and Ar. These results have been described in all details in [8]. Here, we will only concentrate on one particular aspect, which is the relative evolution of the recoil scattering *vs* the binary scattering.

Figure 1 shows (e,2e) triple differential cross sections (TDCS) for the three targets, obtained for a fixed scattered electron observed with an energy E_a of 500 eV at a scattering angle of - 6° with respect to the incident direction, and a sequence of ejected electron energies ranging from 12 to above 200 eV. The vertical thin line indicates the momentum transfer direction and it's opposite. Many various theoretical models were applied (with some success) to describe these experimental data. We only mention the two of them which are needed for the present discussion.

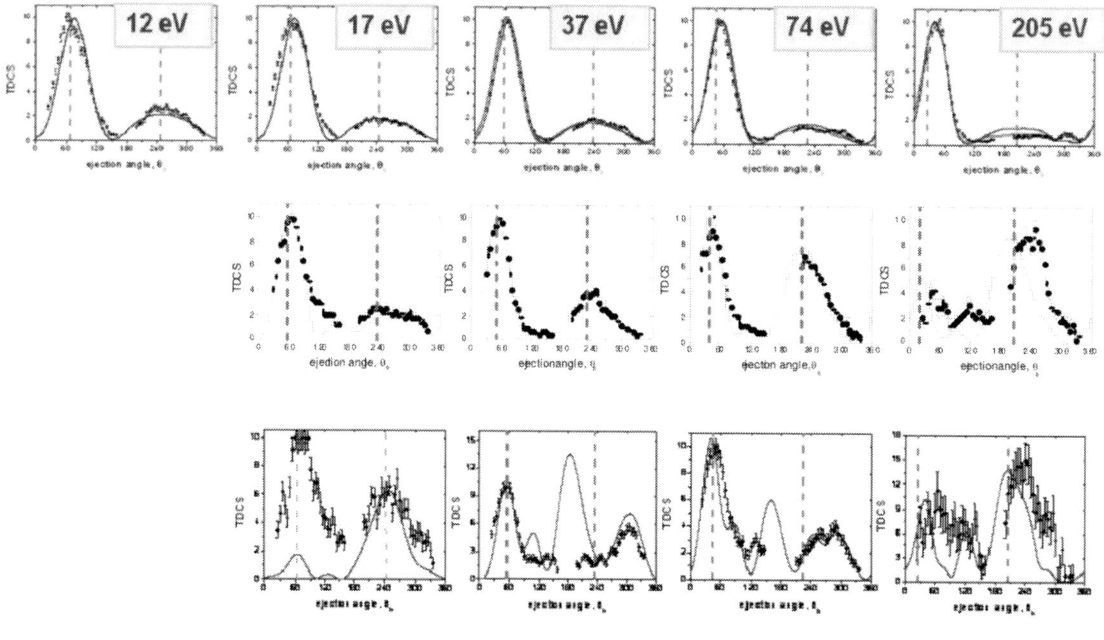

Figure 1. From top row to bottom row: TDCS (given in arbitrary units) for ionization of He ($1s^2$), Ne ($2s^2$) and Ar ($3s^2$), respectively. Blue and red curves represent DWBA-G and CCC calculations, respectively. E_a = 500 eV, θ_a = -6° and E_b as indicated in each column.

First, the Convergent Close Coupling (CCC) model [9] which is known to produce very accurate TDCS results, at least for electron impact SI of He when the residual ion is left in its ground state. Indeed, we find very good agreement between CCC results and our data. Second, the Distorted Wave Born Approximation [8] corrected for the post collision interaction (PCI) by a simple Gamow factor (DWBA-G). At all ejection energies, the CCC (where they exist) as well as the DW calculations are in fairly good agreement with experimental data, except at the largest E_b-value of 205 eV where the ejected and the scattered electrons acquire comparable energies and their PCI becomes strong. Hence, it may not be fully accounted for by the simplified Gamow factor. Also, we observe that the general agreement with experiments deteriorates at all energies from the lighter to the heavier target. For instance, at 37 eV, the theory predicts a prominent central peak positioned in between the binary and recoil peaks, which is not confirmed experimentally. But, besides from these general remarks, what we would like to emphasize from figure 1 is the steady decrease in the case of He of the recoil peak maximum intensity, relative to the magnitude of the binary peak, as the energy of the ejected electron increases. Indeed, the origin of the recoil peak is qualitatively understood as being due to reflection of the ejected electron wave from the atomic potential well. The relative recoil-to-binary intensity ratio, RB, is proportional to the reflection coefficient, which is decreasing as the energy of the ejected electron is getting larger. However, this is not the case for all three targets. For instance, the RB intensity ratio is, in fact, increasing with the ejected electron energy in the case of Neon 2s. Whereas in Argon 3s, we observe that, as the ejected energy increases, the experimental RB ratio is first decreasing, then it becomes minimal at 74 eV and increases again towards larger ejected energies. This tendency is confirmed by the DWBA-G model, except for the lowest energy, 17 eV, where the theory predicts a relatively large recoil lobe in a blunt contradiction to the experiment.

These findings are summarized in Figure 2, where the RB ratios are plotted vs the ejected electron energy for all three targets. The experimental RB ratio for He is decreasing with increasing ejected energy. Conversely, this experimental ratio in Ne is steadily growing across the whole energy range studied here. Whereas the corresponding ratio in Ar is non-monotonous: there is a hint of a minimum in between 37 and 74 eV, and a considerable growth towards larger energies, while the ratio seems to be flattening at the smallest energies of 17 and 12 eV.

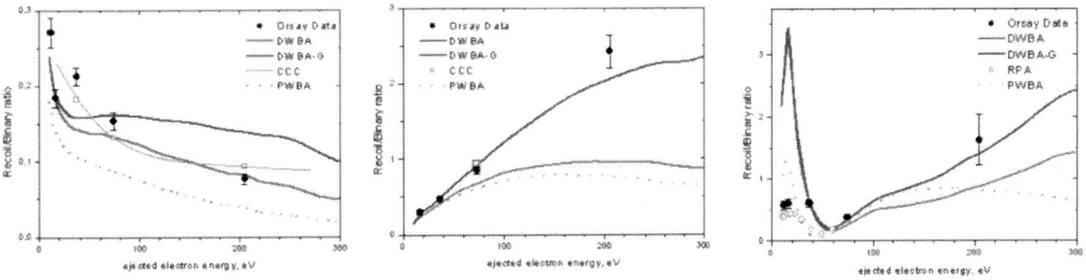

Figure 2. Recoil-to-binary intensity ratio plotted as a function of ejected electron energy for He (left panel), Ne (middle panel) and Ar (right panel).

A theoretical analysis of the different behaviors of the RB ratio was given by Kheifets *et al* [8]. Very briefly, the argument is based on the idea that the RB ratio is determined by the slow electron movement in the ionic potential, which is a superposition of the Coulomb potential of the nucleus and the Hartree-Fock HF potential of the ionized electron core. The smooth, slowly varying Coulomb potential alone is a poor reflector, and it is the short-range HF potential that is chiefly responsible for the formation of the recoil peak. The qualitative change in the RB ratio dependence on the ejected electron-energy from He to Ar can be explained by variation of reflectivity of the short-range HF potential. The reflectivity increases profoundly from He to Ne and to Ar due to specific scattering phase behavior.

3. (e,2e) results on simple molecules

We now turn to (e,2e) results obtained on molecules, and we choose to illustrate this point by the observation of a specific molecular effect, that is the signature of interference effects in the TDCS for the H_2 molecule [10], similar to the Cohen-Fano oscillations predicted for DDCS's [11]. The same idea of a molecular double slit behavior as viewed on electron impact TDCS's was theoretically suggested by Stia *et al* [12]. Very briefly, the idea is the following: we measured (e,2e) angular distributions for H_2 at a number of ejected electron energies, similar to those displayed in figure 1. These H_2 distributions are then compared to those for He. Figure 3 is an example of such comparison.

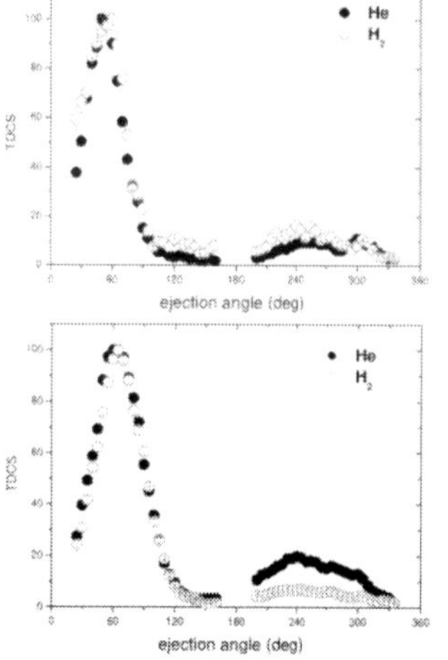

Figure 3. TDCS measured for He (full circles) and H_2 (open circles). E_a = 500 eV, θ_a = -6° and E_b = 205eV (upper panel) and 37 eV (lower panel). The results from the two targets are normalized to each other at the maximum of the binary lobe.

We observe that in one case, at the ejection energy of 37 eV, the recoil intensity in H_2 is diminished with respect to He, whereas it is enhanced at the energy of 205 eV. Such recoil intensity destruction or enhancement in the molecular case was theoretically shown by Stia *et al* [12] to result from Young type interference in which the two sources of coherent emission are the two atomic centers in the molecular target. In fact, Stia *et al* introduced the well known interference factor, I, which reads

$$I = 1 + [\ \sin(q\rho) / q\rho\ ]$$

where ρ is the equilibrium internuclear distance in the molecule and q is the momentum of the recoiling ion. Of course, q depends on the ejected electron energy and hence on its wavelength. Figure 4 shows, for three ejected electron energies, the interference factor, I, derived from our experiments compared to the one predicted by the above formula.

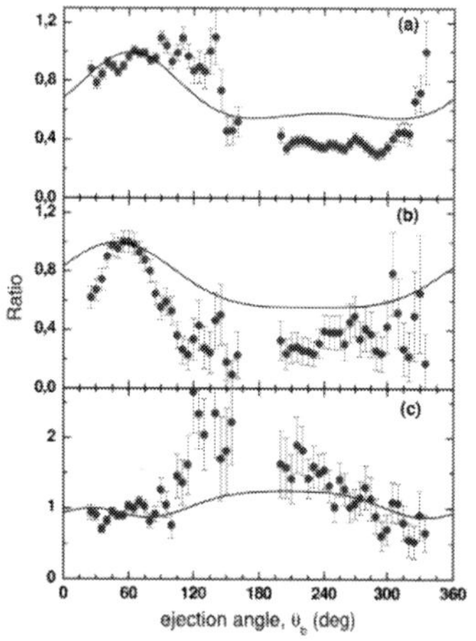

Figure 4. Theoretical interference factor, I, as compared to the experimental ratio of TDCS measured for He and H_2. E_a = 500 eV, θ_a = -6° and E_b = 37, 74 and 205 eV from top to bottom, respectively.

Qualitative good agreement is seen in the three cases between theory and experiments, in spite of the large error bars due to the fact that, in certain angular ranges, we are taking the ratio of two small quantities. These data clearly show a suppression of the recoil peak intensity with respect to the binary one at 37 and 74 eV, whereas the data at E_b = 205 eV display its prominent enhancement. This reasonably good agreement of the experimental results with the predictions of Stia *et al* suggests that the present observations can be ascribed, in the 37 and 74 eV cases, to destructive interference effects arising from the two-centre nature of H_2, or to constructive interferences in the 205 eV case. We note that, to our best knowledge, this is the first time that both the destructive and constructive characters of the interference process are simultaneously observed in the same (e,2e) experiments.

4. (e,3e) results

We now come to discuss some DI results and we choose to illustrate these by (e,3e) experiments on Ar with Auger emission [13]. We investigate here electron impact *indirect* DI events involving first the removal of one fast electron ejected from an inner shell, namely here the Ar-2p shell, followed by an electronic rearrangement which leads to the ejection of one fast Auger electron with an energy 205 eV in this Ar case. We deliberately choose the energies of both continuum electrons to be equal, hence the ejected and the Auger electrons are fully indistinguishable (we do not know which is which!!). This indirect process is in competition with the *direct* DI process involving the removal of 2 electrons from the outer 3p-shell, with identical energies, 205 eV each. Both direct and indirect DI processes yield the same final state, hence they might interfere.

The kinematics of these (e,3e) experiments is as follows. A coplanar geometry is used. The incident electron with energy E_0 = 953 eV is observed after scattering with an energy E_a = 500 eV and under a scattering angle of 6°. The two indistinguishable electrons labeled here b and c for convenience are observed at the same energy in the angular ranges 20 – 160° and 200 – 340°, and all three electrons are detected in triple coincidence.

The data are displayed in figure 5 as a 2-D representation of the measured (e,3e) intensity *vs.* θ_b and θ_c in these angular ranges.

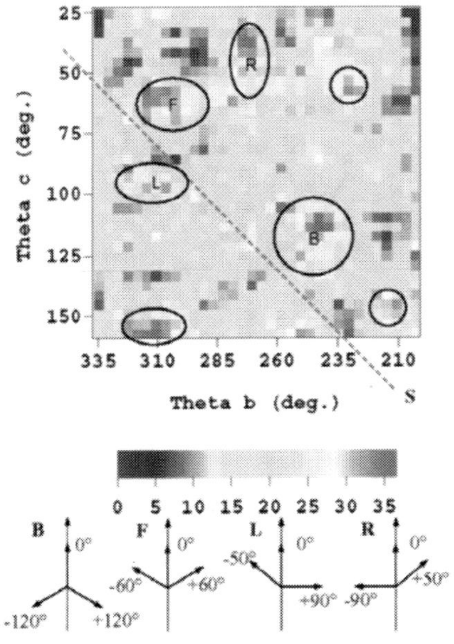

Figure 5. (e,3e) measured angular correlation diagram for two electrons emitted from an Ar target with identical kinetic energy 205 eV, at an impact energy E_0 = 953 eV and a scattering angle θ_a = -6°. The bar gives the relative scale of the cross section. The diagrams at the bottom illustrate the emission directions of the three electrons, for the peaks labeled B, F, R and L, respectively, see text.

The picture is very rich but also very difficult to interpret. Very briefly, we can distinguish several islands (or peaks) of maximum intensity, some of them are labeled B, F, L and R, for which a tentative interpretation is given, associated with the emission diagrams shown at the bottom of the figure, and some others are not labeled.

In peak B, both atomic electrons are emitted backwards (hence B) with respect to the incident direction, at respective angles of 120° and 240° (or –120°). We remind that the fast scattered electron is observed at 6°, i.e. very close to 0°. Hence, the 3 electrons emerge at about 120° from each other, which is the configuration which minimizes the Coulomb repulsion between these final state electrons. In peak F, the two electrons are emitted in the forward direction, at angles of + and – 60°. This observation, coupled with other observations from complimentary (e,3-1e) experiments which are not discussed in the present paper, see [13] for more details, leads us to attributing this peak F to a shake-off (SO) - 2step - indirect DI where the first step is an (e,2e) SI of the 2p-shell followed in a second step by a quasi-isotropic Auger emission.

In peaks L and R, (standing for Left and Right with respect to the incident beam) one electron is emitted close to + 90° in L and close to -90° (or +270°) in R, while the 2nd electron appears at -50° or +50°. These 90° angles are reminiscent of a classical binary collision between two particles of equal mass, hence the interpretation of these peaks as being due to a direct DI in the so-called "Two Step 2" (TS2) process, that is, a binary collision first occurs between the incident electron and one target electron, followed by a 2nd step where the projectile hits a 2nd target electron in an (e,2e) collision. The other peaks, unlabelled, may speculatively be attributed to the interference between the competing direct and indirect DI processes, leading to the same final state.

5. Summary and conclusion

We have discussed in this paper a number of new (e,2e) experiments for ionization of a series of atoms and molecules. For the rare gases, the particular kinematics (characterized by large q-momentum imparted to the recoiling ion) enhances the recoil scattering intensity. Investigating this recoil intensity shows that for the sequence He, Ne, Ar, a different behavior of the RB ratio is observed and it is attributed to the difference in reflectivity of the HF ionic potential. For molecules, we have seen the signature of Young type (destructive or constructive) interference effects, due to the two-centre nature of H_2. We also briefly described (e,3e) experiments involving for the first time coincidence detection of scattered – ejected – Auger electrons, and we have shown that it is possible to disentangle from the data the contribution of different DI mechanisms.

As a final remark we want to add that we badly need further development of theoretical models to help understanding all these very complex SI as well as DI processes.

Acknowledgments

We wish to acknowledge all the colleagues who have contributed to the works quoted in this presentation, namely O Al-Hagan, P Bolognesi, I Bray, F Catoire, O Chuluunbaatar, C Dal Cappello, D V Fursa, B Joulakian, A Kheifets, D H Madison, F Mezdari and X G Ren.

References

[1] Ehrhardt H, Schulz M, Tekaat T and Willmann K 1969 *Phys. Rev. Lett.* **22** 89
[2] Lahmam-Bennani A, Dupré C and Duguet 1989 *Phys. Rev. Lett.* **63** 1582
[3] Lahmam-Bennani A *J. Phys. B: At. Mol. Opt. Phys.* **24** 2401
[4] Weigold E and McCarthy I E 1999 *Electron Momentum Spectroscopy* (Kluwer Academic, Dordrecht/Plenum Publishers, New York)
[5] Lahmam-Bennani A 2002 *J. Electron Spectrosc. Relat. Phenom.* **123** 365
[6] Berakdar J, Lahmam-Bennani A and Dal Cappello C 2003 *Phys. Reports* **374** 91
[7] Catoire F, Staicu Casagrande E M, Lahmam-Bennani A, Duguet A, Naja A, Ren X G, Lohmann B and Avaldi L, 2007 *Rev. Sci. Instrum.* 78 013108
[8] Kheifets A S, Naja A, Staicu Casagrande E M and Lahmam-Bennani A 2009 *J. Phys. B: At. Mol. Opt. Phys.* **42** 165204
[9] Bray I, Fursa D V, Kherifets A and Stelbovics A 2002 *J. Phys. B: At. Mol. Opt. Phys.* **35** R117
[10] Staicu Casagrande E M, Naja A, Mezdari F, Lahmam-Bennani A, Bolognesi P, Joulakian B , Chuluunbaatar O, Al-Hagan O, Madison D H, Fursa D V and Bray I 2008 *J. Phys. B: At. Mol. Opt. Phys.* **41** 025204
[11] Cohen H D and Fano U 1966 *Phys. Rev.* **150** 30
[12] Stia C R, Fojón O A, Weck P F, Hanssen J and Rivarola R D 2003 *J. Phys. B: At. Mol. Opt. Phys.* **36** L257
[13] Naja A, Staicu-Casagrande E M, Ren X G, Catoire F, Lahmam-Bennani A and Dal Cappello C 2007 *J. Phys. B: At. Mol. Opt. Phys.* **40** 2871

Journal of Physics: Conference Series 235 (2010) 012005 doi:10.1088/1742-6596/235/1/012005

Spin-polarised Stokes parameters of open and closed shell transitions and their resonances in zinc atoms.

J F Williams, L Pravica, D Cvejanovic, S Napier, S Samarin and M Piwinski[1]
ARC Centre of Excellence for Antimatter and Matter Studies
School of Physics, The University of Western Australia,
Crawley, Perth. WA 6009. Australia.
[1]Nicolas Copernicus University, Torun, Poland.

E-mail: jfw@physics.uwa.edu.au

Abstract. The formation and decay processes of negative ion resonances are explored for zinc atoms using spin-polarised incident electrons with observation of visible photons from the resonance decay into the 6s ^{1}S, 4d ^{1,3}D and 4p ^{1,3}P states. For Fano resonances, near singly and doubly excited neutral atom states and two ionic states, the Stokes parameters of the radiation indicate markedly different effects of electron exchange and spin-orbit interactions as well as the angular momentum effects of closed and open 3d^{10} core states.

1. Introduction

The advances with atomic clock technology [1], quantum information processing [2], the production of ultracold molecules [3] and quantum phase transitions [4] have aroused new interest in Feshbach resonances with possibilities for control of the interaction strength between atoms [5] and consequently for the study of quantum many-body phenomena. An understanding of the formation and decay of resonances in such applications has followed from the fundamental descriptions [6, 7] and led to more detailed studies of the decisive role played by multi-electron correlations in a wide variety of atoms and collision processes. Photons, rather than electron or other probes, have fewer degrees of freedom and frequently lead to more specific information such as multielectron correlations. For example, in the photo detachment of a 1s electron from a lithium negative ion [8, 9], the rearrangement of the core electrons with various relaxation effects, a strong polarisation interaction between the core and the photoelectron, and a double excitation process in which the post-collision interaction has a dominant role, have been identified in the formation of shape resonances. More detail of angular momentum phenomena has emerged from the photo-absorption studies in Ca and Sr where [10] showed how the spin-dependent interaction affects the mixing of different states and the breakdown of LS-coupling in various auto-ionising series affects the resonance structures. Electron impact studies have extended the observed interactions to include non-dipole, electron exchange and dynamical correlation effects on the formation and decay of auto-ionising resonances [11].

The present paper extends the above knowledge using incident spin-polarised electrons to explore orbital and spin angular momentum through electron exchange and spin-orbit interactions involving negative ion resonances near auto-ionising levels and on their formation and decay processes. The study has approached such phenomena by observing resonance radiative decays, specifically into the 4d, 5d, 6d ^{1,3}D and 4p ^{1,3}P states of neutral zinc atoms in threshold regions using spin-polarised and non-polarized incident electrons [12-19]. Figure 1 shows the studied energy levels and the observed negative ion resonance levels above the ionization threshold which were detected through their decay into the ^{1,3}D and 4p ^{1,3}P states and the subsequent fluorescence with 636.2, 328.2 to 334.6, 277.1 to 280.1, 257.0 to 260.9 ranges and 213.4 nm photons. In this way information could be deduced about the neutral atom radiation with closed shell transitions while the negative ion resonance decays were

Journal of Physics: Conference Series **235** (2010) 012005 doi:10.1088/1742-6596/235/1/012005

observable via open 3d^9-shell formation with subsequent decays via open-to-closed shell paths. These transitions were seeking more tightly bound 3d^{10} core states as expected since there are no ground state atomic states, or singly excited states, with a 3d^9 electronic configuration. In that way we were able to explore a variety of near threshold excitation processes and their associated negative ion resonance states.

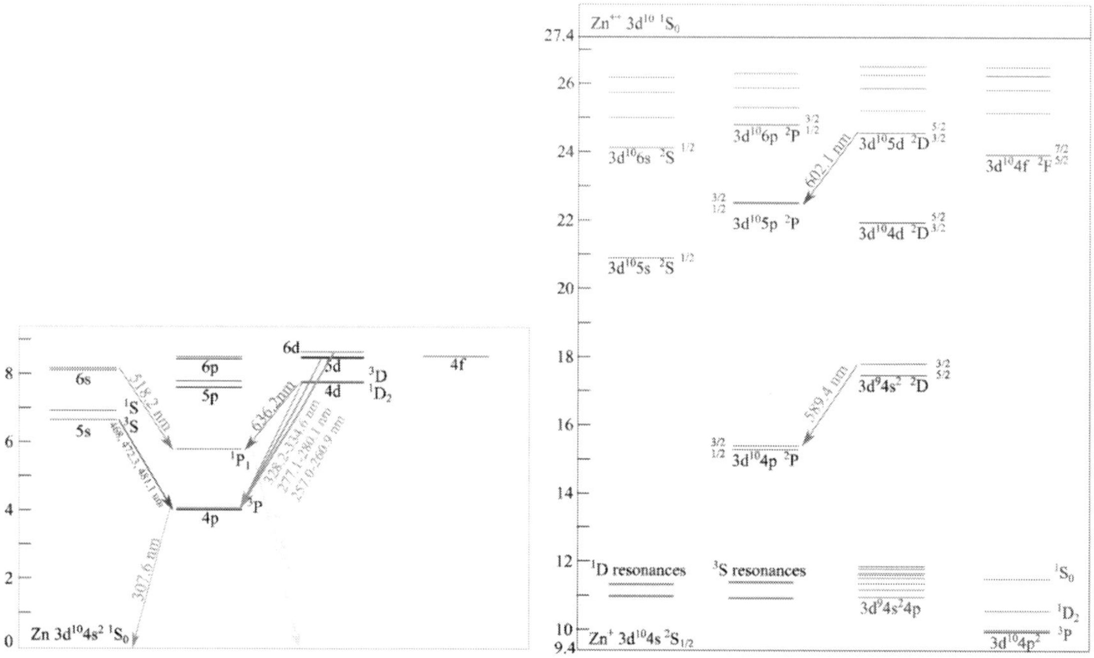

Figure 1. Energy level diagram of zinc atoms, showing the studied levels and transitions of (i) the neutral atom (left hand side) and (ii) the doubly–excited neutral atom, the negative ion and the singly charged positive ion (right hand side).

An underlying approach for this work was the earlier success of deducing the angular momentum exchange from the mixed state electronic configurations of transitions within the neon $3P_{3/2,1/2}$ transitions after excitation using spin-polarised incident electrons and observing the Stokes parameters. It was shown [18], for example, that the polarization of the decay photons depends on the different LS-mixing properties of the intermediately coupled states; the triplet components may make exchange and spin-orbit interaction important while for states with a large singlet component the spin-orbit interaction may be dominant. In particular, to highlight the strength of the technique, it is noted that the 703.3 nm transition ($2p^53p[1/2]_1 \rightarrow 2p^53s[3/2]_2$) contained the most striking feature of a strong resonance observed in the polarised excitation functions at about 18.7 ± 0.2 eV. While the overall magnitude of the linear polarization was quite small, it was the practically zero values of P_1 and P_2 which clearly indicated the influence of the dominant (98.2%) ^3S component of its superposition of LS-based wavefunctions in a P-state. The circular polarisation P_3 was about 0.45 and attributable to the electron exchange process as the dominant excitation mechanism for the state. The small non-zero negative values near -0.002 of the P_2 parameter were attributed most probably to the effects of the negative ion resonances since the spin-orbit coupling effects within the atom is expected to be small for this state. We aimed to see the applicability of aspects of these results to other atoms.

Those experiments led to a number of significant conclusions that in turn indicated the direction for future studies and a subset was selected for study in the present paper.

1) For transitions coming from a common upper state, those with $\Delta J = 1$ had larger linear polarizations than those with $\Delta J = 0$ and -1.
2) For the transitions coming from upper states with a different core state, those with a $^2P_{1/2}$ core had a larger linear polarization than those with a $^2P_{3/2}$ core.
3) The normalized integrated state multipoles Im $T(J)_{11}$, reflecting the electron exchange process, and the Re $T(J)_{21}$ indicating the strength of spin–orbit interaction within the atom, are consistent with calculated LS compositions [20] (Luke 1986).
4) The breakdown of LS-coupling for the $J = 1$ states in the neon 3p manifold was confirmed and hence the spin–orbit interaction within the atom plays a significant role in the electron impact excitation collisions.
5) The negative ion resonances had a significant influence on the polarizations which depended on the different LS mixing properties of the intermediately coupled states.
6) The effect of the core on the alignment of the excited states indicated the resonances have a more significant influence on the alignment of the excited states with a $j = 1/2$ core than for states with a $j = 3/2$ core. The state multipoles indicated the distribution of the integrated charge cloud of the excited states and the angular momentum transferred during the collision.

Results 1) and 2) indicated the need for consideration of the angular momentum coupling in the excitation processes. Results 3) and 4) indicated the advantages of modeling to indicate probable correlated wave functions and their mixing coefficients. Results 5) and 6) indicated the angular momentum coupling in the excitation process, the importance of observing the polarization of the transition radiation as well as the need for modeling and for angular correlation measurements,

Questions remain about the propensity for similar transitions which we pursue here for zinc atoms with observations of closed and open shell transitions. Information of the angular momentum transfer and its association with the multiple moments of the excited state will be pursued with the observation of the "integrated" (over the angular distribution of the scattered electrons) Stokes parameters P_i (i= 1, 2, 3) of the excited states. We report the Stokes parameters for the 6s ^1S to 4p^1P transition to establish the precision of the technique and to verify its applicability to zinc. Then we report the P_i for the 4d ^1D transitions to establish the closed shell behaviour below the first ionisation level, then explore the excitation of the ionic closed shell $3d^{10}5d$ ^2D to $3d^{10}5p$ ^2P 602.1 nm transition and the open shell $3d^94s^2$ ^2D to $3d^{10}4p$ ^2P 589.4 nm transition to explore the spin-orbit and exchange interactions. Finally the observations of open and closed shell transitions in the vicinity of the resonances in the autoionizing states near 11 eV are explored to complement the energy loss electron studies [15].

2. Apparatus and techniques

The observations follow from the use of incident low energy spin-polarised electrons to control the spin angular momentum; a beam of thermal energy zinc atoms to define the initial ground state; scattered electron energy loss and radiated photon detection to identify the scattering state and polarization detection to identify the angular momentum transfer in the scattering process as well to identify electron exchange and spin-orbit interaction effects.

The apparatus is described in detail elsewhere [13] and briefly here. The intersection of crossed electron and zinc atom beams is observed by photon polarization detectors and electron energy analysers. Either unpolarized or polarized incident electron beams may be used, the latter produced via 830 nm photo emission from a GaAs surface after oxygen and cesium treatment and energy selected by an electrostatic 180° hemispherical monochrometer. The incident electron beam was about 100nA with 30% polarization. Base pressures are about 4×10^{-11} Torr in the spin-polarised source region and about 5×10^{-8} in the zinc interaction chamber. Optical interference filters preceded the photomultiplier detectors and, for example, selected the 518.2 nm photons from the 6s ^1S state with a 518.3 nm central wavelength filter with a FWHM of 0.3 nm. The zinc oven reached a stable operating mode when its temperature reached about 410 °C after 1 hour working time when the 468.1 nm photon count rate was about 6000/sec. For these conditions there was no detectable instrumental or collisional depolarization. The observed Stokes parameters have been corrected for hyperfine depolarization arising from the natural abundances of 95.9% and 4.1% for the isotopes with nuclear spin of 0 and 5/2, resp. The scattered electron energy resolution was selected from 150 to 300 meV according to the need of the measurement.

The scattered electrons were detected after energy selection by a 127° cylindrical analyzer. Incident electron energy and also scattered electron residual energy (to maintain a constant energy loss) were scanned simultaneously so that energy dependence of angle differential scattered electron intensity was measured. The scattered electron energy analyser can rotate on both sides of the incident electron beam direction to verify symmetric scattering around the incident beam direction and enable left/right scattering asymmetries. Orientation of the electron spin vector perpendicular (up/down) to the electron scattering plane was changed by reversing the helicity of the laser light in the source.

3. Results and Discussion

3.1 The closed core 6s ^1S to 4p^1P 518.2 nm transition

The precision of the technique was indicated by measuring the Stokes parameters $P_{1,2,3}$ of the 6s ^{1}S to 4p^{1}P 518.2 nm radiation as shown in figure 2.

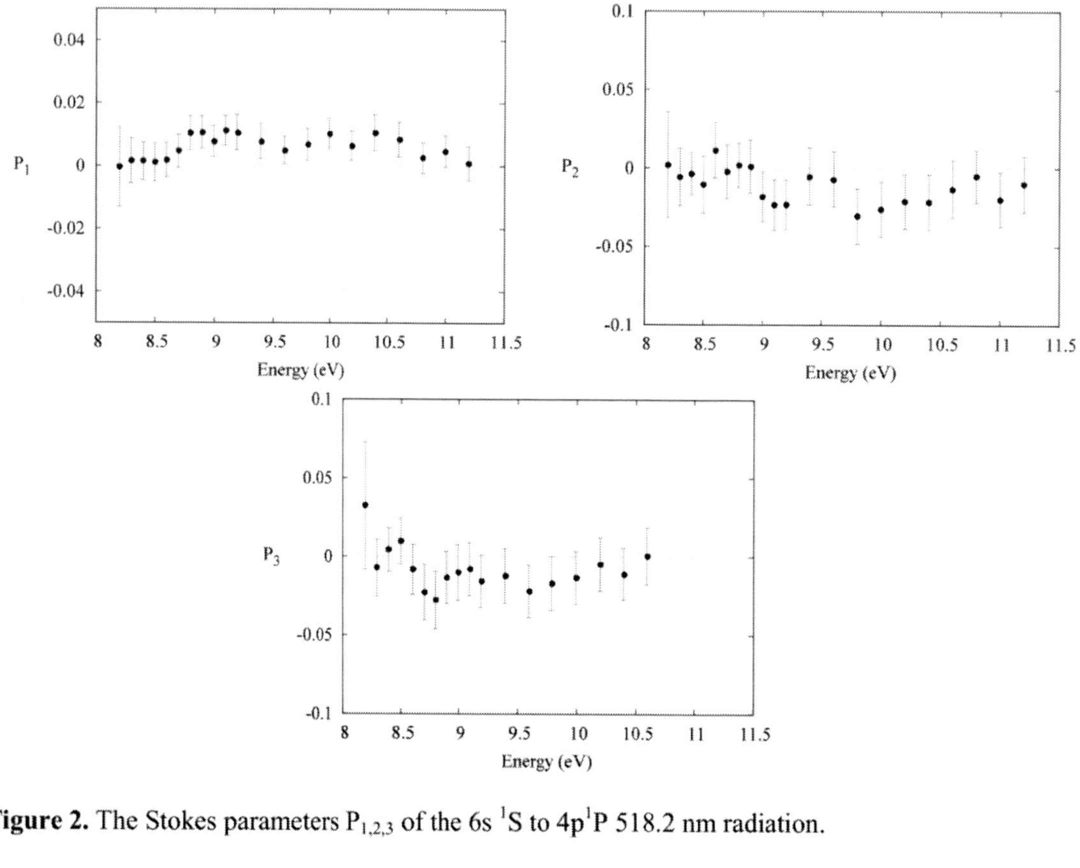

Figure 2. The Stokes parameters $P_{1,2,3}$ of the 6s ^{1}S to 4p^{1}P 518.2 nm radiation.

Within about the first 0.6 eV from threshold all three integrated Stokes parameters (integrated over all electron scattering angles) are zero within the statistical uncertainties. This transition is very weak compared with the other transitions observed in this work and so its measurement established the precision of our technique for the present experimental conditions. Also the observation is consistent with the state being a pure and well LS-coupled state, i.e. the electron charge cloud is spherical, both P_1 and P_2 should be zero and, since the upper state is a singlet state, P_3 is expected to be zero. Those observations indicate the applicability of the interpretation of the measurements to zinc, at least to those states. About 1 eV above threshold the three P_i parameters are definitely non-zero, most likely by transfer of polarization through cascade from higher states, but the possibility of resonance phenomena can not be excluded for the present statistics and limiting electron energy resolution of about 0.3 eV.

3.2 The closed core 4d $^1D_2$ to 4p $^1P_1$ 636.2 nm transition

The measurements of the integrated Stokes parameters for the 4d $^1D_2$ to 4p $^1P_1$ 636.2 nm transition, shown in figure 3, also indicate very small (<0.002) values of P_2 within about 2 eV from the threshold at 7.74 eV and small, near zero, values of P_3. Again the statistical accuracy is sufficiently good to indicate the minimization and/or absence of most experimental uncertainties.

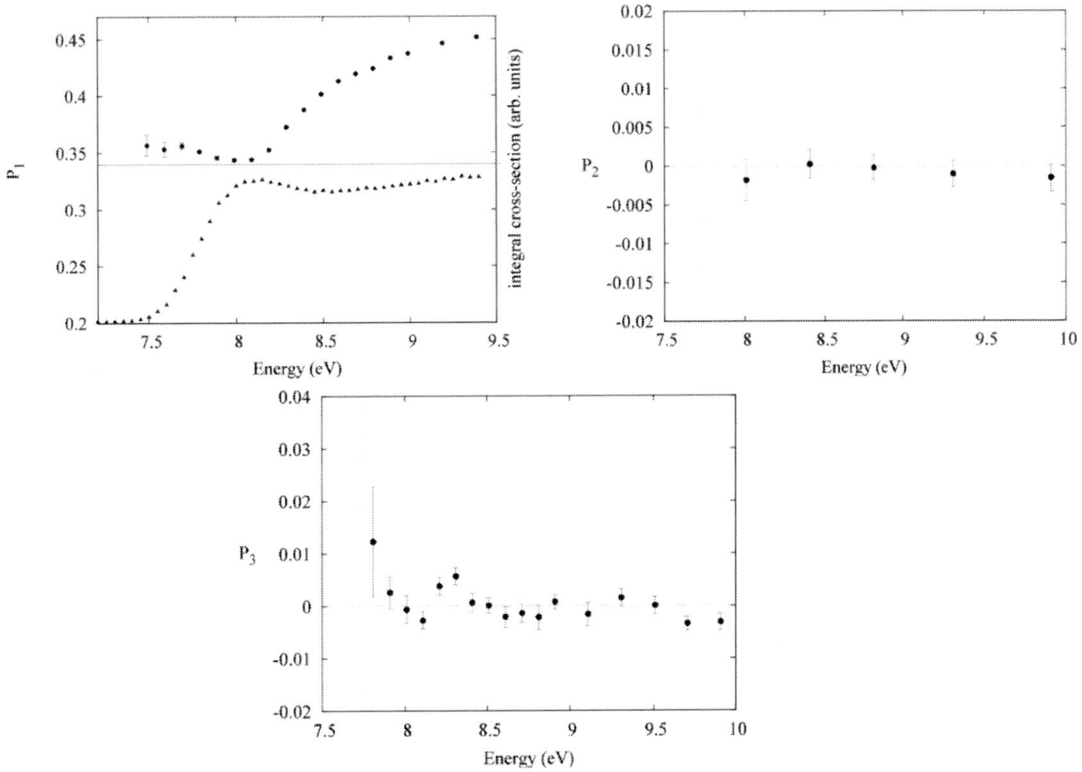

Figure 3. Integrated Stokes parameters $P_{1,2,3}$ for the 4d $^1D_2$ to 4p $^1P_1$ 636.2 nm transition. The vertical line indicates the threshold energy of 7.74 eV. The top left section shows P_1 (upper data) and the integral cross section (lower data, right hand side axis).

The integral cross section for the 4d $^1D_2$ state indicates that cascade from the 5p level is evident from its threshold at 8.2 eV and it contributes significantly to P_1 and possibly with a very small contribution near 8.3 eV to P_3 values. The model of Bartschat and Blum [21] indicates that the spin-orbit interaction is negligible and that the state is well LS-coupled. If the small P_3 values are considered as zero within ± 0.005 the experimental and theoretical indications that electron exchange and spin-orbit interaction are negligible, are in agreement.

3.3 *Open and closed shell transitions near 24 and 18 eV*

The above data for closed core transitions indicate a consistent model for excitation of S and D states of low energy and near threshold. To explore the difference between open and closed shell transitions we choose the ionic $3d^{10}5d\ ^2D_{3/2}$ closed–to-closed $3d^{10}5p\ ^2P_{1/2}$ shell transition and the $3d^94s^2\ ^2D_{3/2}$ open–to-closed $3d^{10}4p\ ^2P_{1/2}$ shell transition, i.e with both parent states well above the first ionisation threshold. Wavelength filters of 589.6 nm and 602.2 nm with 0.6 and 1.0 nm FWHM bandwidths, resp. were used. By the nature of the open shell binding of electrons in zinc, all of the open shell states lie above the first ionisation potential and so there are competing auto-ionisation continua which influence the interpretations. Figure 4 indicates slight differences between open and closed shell transitions in all three Stokes parameters.

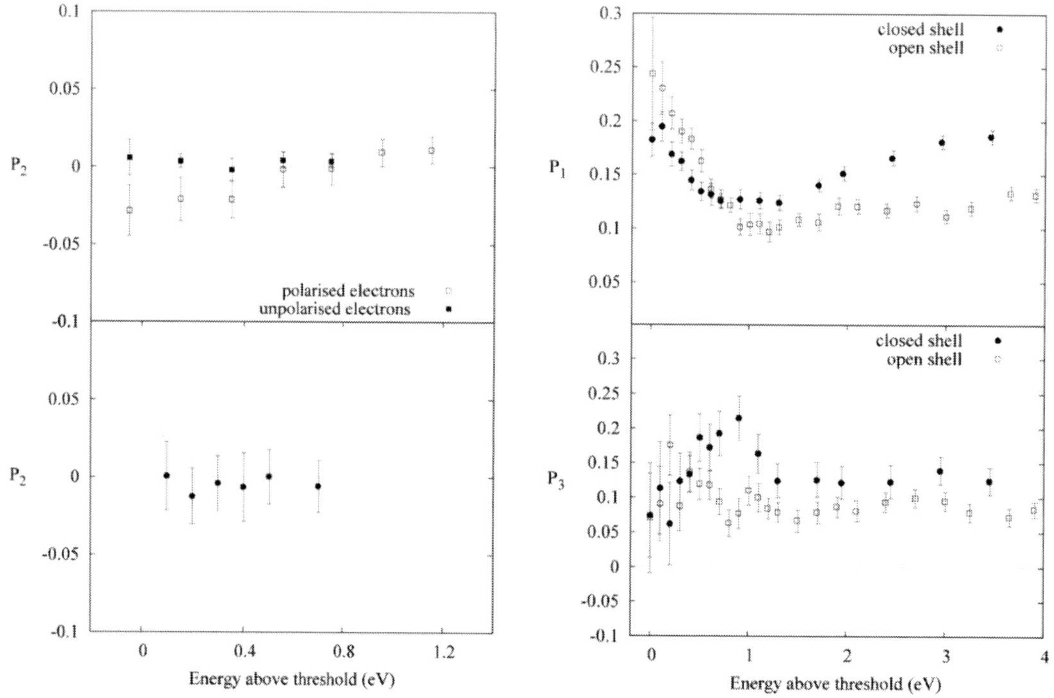

Figure 4.The Stokes parameters P_1 (lower section), P_2 and P_3 (upper sections) for the closed $3d^{10}5d\ ^2D_{3/2}$ to closed $3d^{10}5p\ ^2P$ shell 589.6 nm transition (closed circles) and open $3d^94s^2\ ^2D_{3/2}$ to closed $3d^{10}4p\ ^2P$ shell 602.2 nm transition (open squares).

The top left section of figure 4 shows zero values of P_2 for an incident non-polarised electron beam because of the cylindrical symmetry of the scattering geometry when there is not a spin vector to define a plane. That result confirms a correct working of that aspect of the method as well as the size and spread of the uncertainties in the data points indicating the precision and accuracy of the measurements but only for that observed wavelength. Consequently even the three data points just above threshold indicate non-zero values of P_2. We see that the open shell transition with non-zero P_i values indicated angular momentum transfer with combined exchange and spin-orbit interactions, at least as indicated by this technique, as well as the rearrangement of two electrons; clearly a consequence of the open $3d^9$ core. An interpretation may be that the 3d core-perturbed potential enhances the probability of capturing, albeit temporarily, an electron to form a resonance just above threshold. In contrast the closed shell radiating the 589.6 nm photons showed a zero value of P_2. Both the open and closed shell transitions occur in the process of excitation-with-ionisation, the ejected electron escaping with unknown angular momentum and limiting the present interpretation.

The Stokes P_1 parameters for the open and closed shell transitions within the zinc ion are rather similar with the essential similarities of a sharp decrease in value near threshold presumably indicating a rapid change of the electron exchange interaction shown for P_3 which also appears then in P_1. Assuming in the limiting case at threshold, and at least below the onset of radiation cascade, that only the L = 0 partial wave is dominant a finite polarization occurs at threshold, electron exchange is the dominant excitation process and the decrease of the polarisation with increasing energy occurs.

These measurements have shown how an open shell in the ionization-with-excitation process caused a breakdown of LS-coupling while a closed shell process remained well LS-coupled. In both processes the residual ions remained aligned and oriented. An analysis [13] using normalized state multipoles revealed much more information. For the closed shell transition the total and spin angular momentum transfers, and hence the magnetic dipole moment of the spin angular momentum transfers and the electric quadrupole moment of the orbital angular momentum, were separated. A positive value of the spin magnetic dipole indicated positive angular momentum transfer from the incident electron via exchange. The negative value of the total magnetic dipole indicated a negative angular momentum transfer to the residual ion when the spin coupled to the orbital angular momentum. In that way the m = -1 states were populated rather than the m = +1 states.

3.4 Open core resonances near 11 eV

To seek further evidence of electron correlations we looked with the sensitivity of the Stokes parameters to the region where the $3d^{10}$ core opens just above the first ionization threshold, as indicated in figure 5.

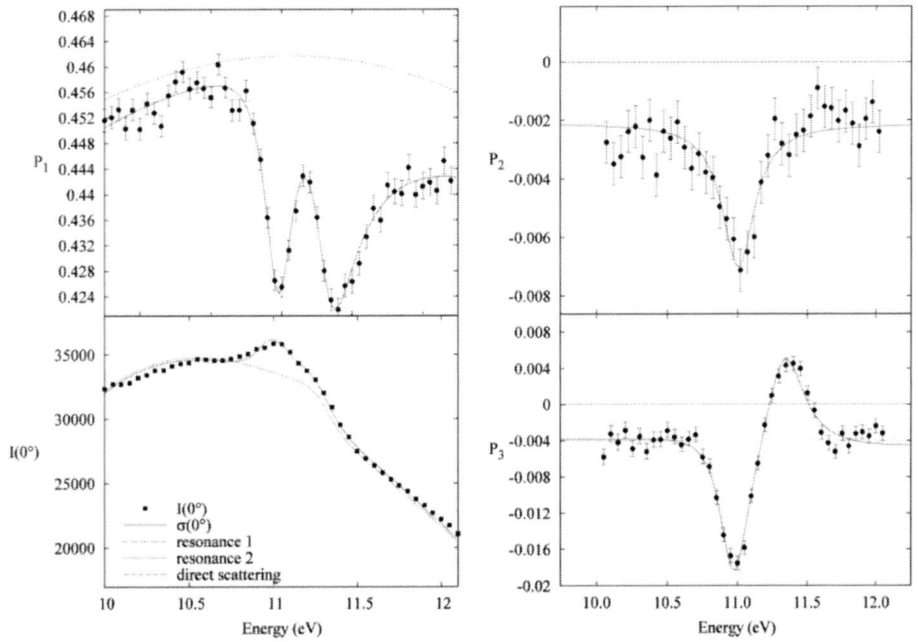

Figure 5. Stokes parameters P_i fitted with Fano resonance profiles for the 4d $^1D_2$ state decay photons. The bottom left section indicates the measured intensity of the linear polarization at an angle of zero degrees. The dashed line indicates the fitted background for the extracted resonances.

In the excitation of the 4d $^1D_2$ state around 11 eV the intensity of the radiated 636.2 nm photons, for either spin-up or spin-down incident electrons, indicates only a localized rise as shown in figure 5, however their difference, i.e. the Stokes parameters, shows not one, but two, resonances.

The lower energy resonance with a significant P_2 value infers a strong spin-orbit interaction in contrast to the negligible (near zero P_2) spin-orbit interaction for the higher-energy resonance. By definition, these resonances are a measure of the configuration interaction between each resonance state and its accessible continuum states. For an isolated resonance the Fano profile shapes $\sigma(E)$ are described by

$$\sigma(E) = c(E) + \{2\sigma_a\, q\, (\Gamma/2)\, (E-E_r) + \sigma_a\, (q^2-1)\, (\Gamma/2)^2\} \,/\, \{(E-E_r)^2 + (\Gamma/2)^2\} \qquad (1)$$

where c(E) is the sum of the resonance σ_a and other σ_b backgrounds, q is the profile index and Γ is the resonance width. In the usual fitting procedure, the two resonances were shown to have energies of 10.98 ± 0.02 eV and 11.33 ± 0.02 eV with widths of 0.25 ± 0.03 eV and 0.33 ± 0.05 eV respectively. Ideally an experiment must determine also the phases of the overlapping scattering matrix elements as well as the differential inelastic cross sections into the accessible channels in order to characterize a resonance completely but while this is not yet feasible some details may be deduced. It was assumed that the two resonances were coupled to different continua, their separate shapes just superposed and no interference between them occurs [22]. Given the width of the resonance, the q-parameter measures the ratio between the transition probabilities to the bound and continuum components of the final state from the initial state. The q value is large for resonances dominated by the contribution from the transition to the bound component of the final state. Here, the q-values are given as (a:P_i) for resonances (a) and (b) with Stokes parameter P(i = 1, 2, 3) are (a: 4.9, 4.4, 4.2) and (b: -2.6, -2.1, -2.4) respectively. Singly-excited closed-core-excitation channels have large q-values and generally reduce the background contributions to the double excitation resonances which, in turn, causes the raising of the q-parameters for the double excitation resonances and which become more symmetric. Here the main core excitations arise from the 3d electrons and the correlations for the 3p electrons are small; the q-values are small and the different signs indicate different phases of the continuum background. The observation is in accord with photo-ionization data which indicate that core-shielding effects on resonance energies and widths are small and in contrast core shielding effects on the q parameter of the resonance are large. Work is in progress to determine the cross sections for the background continua.

Further interpretation of the negative ion resonance data requires a configuration, a coupling scheme and consideration of the effects of electron exchange and the spin-orbit interaction, as follows. Assuming that the additional electron is bound with a negative electron affinity of up to 0.5 eV, as for a Feshbach resonance, then the observed negative ion resonances at 10.98 and 11.33 eV, can be associated with a parent configuration of $3d^9\,4s^24p$ with a calculated [23] center-of-gravity energy of 11.459 eV rather than the lower $3d^{10}4p^2$ configuration at 10.301 eV. A resonance configuration with two outer 4p electrons bound to the ionic grandparent $3d^9\,4s^2$ with J = 5/2 and J = 3/2 ion cores, resp., is probable since most of the $3d^{10}4p^2$ configurations have energies too low and the $3d^9\,4s^25p$ etc energies are too high. The spin-orbit splitting in this grandparent ion core is 0.337 eV [24] in accord with the energy difference of 0.36 ± 0.02 eV for the two observed resonances.

3.5. Multi-configuration Hartree-Fock [MCHF] models

Further assistance with an interpretation of these results for excitation of the 4d ^1D state and the two ^2D states was sought in a multi-configuration Hartree-Fock [MCHF] type calculation [25]. The approach was to achieve an outcome similar to that for neon [18], that is, to treat the closed core with either selective inclusion of linear combinations of configurations or “switching off” different parts of the program to obtain an indication of the effects of the closed and open cores. Variations included (i) closing the 1s-3d shells to allow only the effects of correlation between valence electrons to contribute to the calculation, (ii) remove one 3d electron from the core to allow correlation between core and valence electrons, (iii) remove a second 3d electron to allow for additional core-core correlations, and (iv) allow for relativistic (spin) effects.

The procedure was standard for an outer two-electron closed shell atom, i.e. diagonalise the non-relativistic $H_{HF} = T_1 + T_2 - 2/r_1 - 2/r_2 + 1/r_{12}$ to obtain eigenstates $|i\ ^{2S+1}L> = \Sigma\, c_{ia}\ |a_a\,^{2S+1}L>$ of appropriate symmetry. Add spin-orbit interaction $V_{SO} = \Sigma\, \ell_i.s_i / r^2$ and repeat diagonalisation since V_{so} may cause mixing of multiplets in the representation of the additional eigenstates. This procedure was simple when the 3d electrons were part of the core and the only possible states were 4d, 5d etc and the orbitals were spectroscopic i.e the diagonal energy is related closely to a binding energy and the orbital was basically Hartree-Fock. In contrast, the correlation orbitals can be considered as corrections to the wave function with a large diagonal energy parameter which effectively contracts the orbital and so modifies the wavefunction in the region of the spectroscopic orbitals of the main configurations. The calculations for the 4d ^1D level in the MCHF package were continued until the length and velocity forms of the matrix elements agreed to within 1.3 % and the lifetime and energy agreed with standard values. At that point the core included n = 1, 2, 3 except when one 3d-electron was included in the configuration state functions (CSF) to model the core-valence correlation and n = 7 was the highest principal quantum number of the electron CFSs included.

A similar approach was used for the observed ^2D ionic configurations. When the $3d^{10}$ electrons were part of the core, the only possible states were 4d, 5d etc and the calculations were simple with a small number of CSFs. The correlation orbitals were considered as corrections to the total wave function, they usually had a large diagonal energy parameter which had the effect of contracting the orbital and so modify the wave function in the space occupied by the spectroscopic orbitals of the main configuration. However the lowest eigenvalue for the ^2D ionic configuration is the open core $3d^94s^2$ state, which plays an important role for the higher $3d^{10}$ 4d and 5d ^2D states and so must have good radial functions for the same $3d^{10}$ potential. This was achieved using a Hartree-Fock calculation with an average energy of the two states represented as the average configuration $3d^{10}\ 4d^{0.5}\ 5d^{0.5}$ and then estimating the 4s from the lowest state $3d^9\ 4s^2$ by an MCHF calculation in which only the 4s-orbital was varied. A reasonable estimate of the core-orbitals (1s-3p) and of the 3d, 4s, 4d and 5d orbitals was obtained. The state $3d^94s^2\ ^2D_{5/2,3/2}$ was represented by a single configuration wave function composed, for example, of fixed orbitals 1s, 2s, 3s,4s; 2p,3p, 4p and 3d, 4d, 5d [26]. Iterative processes, such as indicated above, enabled accurate energies for the 4d and 5d ^2D states while the $3d^94s^2\ ^2D$ energy varied depending upon the iterative procedure.

The results of these procedures determined that the valence correlations were limited to the correlations between only a few CSFs. Inclusion of the core-valence correlations made a significant difference and brought the calculated energies to less than 1% to the experimental values and the length and velocity values were almost identical with only a 0.1% difference. For example, the core-valence correlation improved the calculated values of wavelengths by 15% and 8% for the 4d $^1D_2$ and closed shell ^2D transitions respectively showing its significance. However, the calculations did not show any significant configuration mixing of the ^1D and ^2D with the non ^1D and ^2D configurations, respectivelly.

It is interesting to note that the lifetime of the open-shell $3d^94s^2\ ^2D_{3/2}$ state via 589 nm decay is 1870 nsec and it involves the transition of two electrons after the ionisation i.e. of $4s^2$ to 3d and 4p and hence expected to be weak. On the other hand, the closed shell transition involves electrons in the n=5 orbital which is expected to be weaker than the similar n = 4 transitions. The NIST database indicates equal line strengths, previous measurements indicated that at 50 eV the closed shell transition is about fifty times stronger whereas the present measurements [27] indicate, close to the threshold of the two states is about three times stronger. The differences arises from the different excitation functions for the $3d^9\ 4s^2\ ^2D$ and $3d^{10}$ 4d ^2D transitions for which it appears that the $3d^9\ 4s^2\ ^2D$ state is formed by the removal of a single inner-shell electron while the $3d^{10}$ 4d ^2D state is formed by ionization of one electron and excitation of another electron.

Consequently, the non-zero values of P_2 for the open-shell transition may be attributed to configuration mixing in the 3d orbital combined with spin-orbit interaction effects. This result is consistent with findings for ejected electron spectra [28] and photo-ionisation [29]. However it is the spin-polarised effects that are the more sensitive test of the interplay of correlation effects with exchange and spin-orbit interactions because of the observations when the strength of those interactions is not so small compared with the Coulomb interaction and the relative (unknown) phases of the interfering wavefunctions. Similar effects are found in various electron scattering phenomena to occur when the Coulomb dominated cross section is minimal and their spin effects become observable. Further work is needed to quantify the wavefunctions and their phases for the open and closed channel resonance formation in the open- and closed-core states in the region above the first ionization threshold, particularly for overlapping resonances.

5 The spin-dependent angle differential scattering asymmetries for resonance states

Further details and confirmation of the electronic configurations of the above resonances in section 3.4 observed in the 4d ^1D photon decay channel near 11 eV await measurements of their spin asymmetries. However experimental considerations have limited observation of spin asymmetries to the 4s4p ^3P electron energy loss channel which are reported here. Figure 6 shows the spin dependent

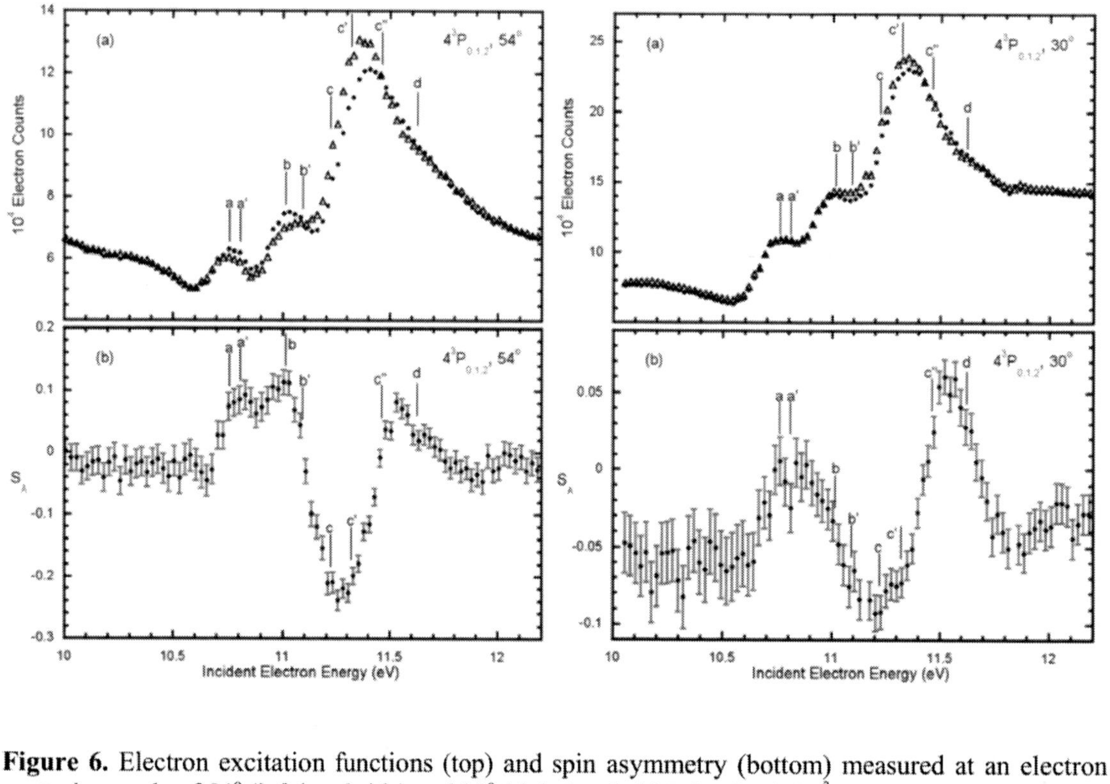

Figure 6. Electron excitation functions (top) and spin asymmetry (bottom) measured at an electron scattering angle of 54° (left hand side) and 30° (right hand side) for the 4s4p $^3P_{0,1,2}$ states of zinc atoms. The data for the excitation functions indicate spin-up (triangle) and spin-down (squares) and their difference, normalized to their sum and the incident electron beam spin polarization, is the spin asymmetry. The short vertical lines and the labels a, a' etc indicate the energies and identify the resonances from other measurements (see text).

angular differential scattering intensities for the 4s4p $^3P_{0,1,2}$ states, and their corresponding spin asymmetry functions, for electron scattering angles of 30 deg and 54 degrees over the energy range from 10 to 12.2 eV. The spin asymmetry is defined as the difference of the scattered electron intensities for incident electron spin-up and spin-down, normalized to their sum and the incident electron beam spin polarization. An initial understanding of the features in figure 6 is derived from various sources.

The present measured electron energy scale was determined to within ± 0.030eV from the onset of radiation from the 4d ^1D state at 7.740 eV. The present electron energy resolution was not sufficient to separate the resonance energies. However their identification with configurations $3d^9\ 4s^24p4d$ was obtained from further observations of photon emission from the 4p $^{1,3}P$ states [17], the 4d,5d and 6d $^{1,3}D$ states [16] and of elastic electron scattering [19]. The negative-ion resonance structures are indicated by the labels a, a', b, b', c; c' and c'' and d with their (energies and, where available, widths) in brackets, as follows: a(10.74), a'(10.82), b(11.01, 0.25), b'(11.10), c(11.22), c'(11.33, 0.33), c''(11.46) and d(11.61) eV. The energies for resonances labeled a, a' and b, b', are determined from the derivative of transmitted current [19], the energies of c, c' and c'' are the average of peak energies observed in photon emission [16], and the energy of the structure labeled d comes from differential elastic scattering [19]. The differential elastic scattering [19] indicated that all negative ions in the $3d^94s^24p^2$ manifold are formed from, and decay by, emission of a d-wave electron. The two lowest energy structures labeled a, a' and b, b' were clearly doublets.

The energy dependence of the spin asymmetries shows marked differences between 30 and 54 degrees electron scattering angles. Both above and below the resonance region, the 30 deg data show a small but statistically significant non-zero asymmetry in contrast to the zero values for the 54 degree data which suggests a d-wave nature of the 4s4p direct excitation process and then presumably also through the resonance regions. A similar behaviour exists for the angle integrated Stokes parameters P_1, P_2 and P_3 of the singlet $3d^{10}$ 4s4d $^1D_2$ and $3d^{10}$ 4s4d $^1P_1$ decay photons [16] that both spin-orbit and exchange are active in the observed $3d^9\ 4s^2\ 4p^2$ negative ion resonances. Through the resonance regions there are significant interferences between the resonant process and the direct excitation processes.

Further interpretation is restricted by the difficulty of the measurements and the lack of data from other sources. A relationship between resonance decay width and spin alignment may be inferred, for instance from helium-like systems [30] where the widths of the singlet resonances are larger than those of the corresponding triplet resonances. Also for triply excited states, the spin alignment determines the relative branching ratios of the decay channels. For example, the branching ratios of the $2s2s2p^2P_0$ state of lithium, the two 2s electrons are most likely to interact strongly and the 2s electron with spin anti-parallel to that of 2p is most likely to escape and a similar conclusion can be drawn from the decay of other triply excited states in three-electron or four-electron systems. The present data indicate real expectations of resolving the resonances and their spin asymmetries with achievable electron energy resolutions. However there are insufficient data presently to extract the direct excitation phase shifts. Further measurements are in progress.

4. Conclusions

A wide variety of measurements have been made of the zinc excitation processes. Observations of the "integrated" Stokes parameters P_i, i.e. integrated over the angular distribution of the scattered electrons, of the photons radiated from polarized electron excited closed $3d^{10}5d\ ^2D_{3/2}$ and open $3d^94s^2$ $^2D_{3/2}$ shell ionic states of zinc, as well as the 6s ^1S and 4d ^1D state of neutral zinc atoms, have been made. The data describe resonance and non-resonance behaviour as well as the effects of electron exchange and spin-orbit interactions on the excitation and excitation-with-ionisation processes.

The measurements have shown how an open-core in the ionization-with-excitation process caused a breakdown of LS-coupling while the closed-core process remained well LS-coupled. For the latter case using state multipole analysis it was possible to separate orbital and spin angular momentum transfers to be separated and show the sign of the angular momentum transfer to the residual ion and then the preference for populating different magnetic sublevels. Measurements of the spin up / down asymmetries for inelastically scattered electrons indicated relatively large effects from the resonances observed in the 4s4p $^3P_{0,1,2}$ differential excitation probability. Initial descriptions of the energies, widths and electronic configurations of the resonances have been made. The significant value of spin polarised studies of the 3d transition metal atom zinc in revealing quantum structure and scattering dynamics has been shown.

Acknowledgements
This work was funded in part by the Australian Research Council Centre of Excellence program and by the University of Western Australia. SN was supported by an Australian Postgraduate Award. MP received an Australian Government Endeavour Research Award.

References
[1] Boyd M et al 2006 *Science* **314** 1430
[2] Hayes D, Julienne P S and Deutsch I H 2007 *Phys Rev Lett* **98** 070501
[3] Regal C A et al. 2003 *Nature London* **424** 47
[4] Bourdel T et al *2004 Phys Rev Lett* **93** 050401
[5] Chung Kwong T 2004 *Rad Phys Chem* **70** 83
[6] Burke P G 1965 *Advan. Phys.* **14** 521
[7] Chen JCY 1970 *Nucl. Instrum. and Meth.* **90** 237
[8] Ivanov V K and Yatsyshin P I 2009 *Tech Phys* **54** 7
[9] Lindroth E and Sanz-Vicario J 2004 Rad Phys Chem **70** 387
[10] Chen J H, Fang T K, Chu C C, Yuh T S and Chang T N 2009 *Phys. Rev.* A **80** 042509
[11] Watanabe T, Domondson A and Tong X 2009 *Phys. Rev.* A **80** 042709
[12] Napier S N, Cvejanovic D, Williams J F and Pravica L 2007 J. *Phys. B: At. Mol. Opt. Phys.* **40** 1323
[13] Pravica L, Williams J F , Cvejanovic D and Napier S A 2007 *Phys Rev* A **75** 012721
[14] Pravica L, Cvejanovic D, Williams J F and Napier S A 2007 *Phys Rev* A **75** 030701R
[15] Napier S A, Cvejanović D, Williams J F and Pravica L 2008 *Phys Rev* A **78** 022702
[16] Napier S A, Cvejanovic D, Williams J F and Pravica L 2002 *Phys Rev* A **78** 032706
[17] Napier S A, Cvejanovic D, Williams J F, Pravica L, Fursa D, Bray I, Zatsarinny O and Bartschat K 2009 *Phys Rev* A **79** 042702
[18] Yu D H, Cvejanovic D, Williams J F , Pravica L, Srivastava R, Stauffer A, Hayes P A, and Napier S 2006 *AIP Conf Proc* **811** 185. Eds Lahmam-Bennani and Lohmann B
[19] Napier S A, Cvejanovic D, Burrow P D, Williams J F, Michejda J A and Pravica L 2009 *Phys. Rev.* A **80** 042710
[20] Luke T M 1986 J. *Phys. B: At. Mol. Opt. Phys.***19** 843
[21] Bartschat K and Blum K 1982 *Z Phyz* A **304** 85
[22] Mies FH 1968 Phys Rev **175** 164
[23] Mansfield M W D 1981 J. *Phys. B: At. Mol. Opt. Phys.***14** 2781
[24] Ralcchenko Y 2005 *Atom Spect Database* (ver.3.0.1) Online http://physics.nist.gov/asd3
[25] Froese Fisher C 2000 *Comput. Phys. Commun.***128** *635*
[26] Glass, R. and Hibbert, A 1978 *Comput. Phys. Commun.* **16** 19
[27] Inaba S, Hane K and Goto T 1986 *J. Phys. B: At. Mol. Opt. Phys.***19** 1371
[28] Back C G, White M D, Pejcev V and Ross K J 1981 *J. Phys. B: At. Mol. Opt. Phys.***14** 1497
[29] Bartschat K 1987 *J. Phys. B: At. Mol. Opt. Phys.***20** 5023
[30] Chung K T 1998 *Phys Rev* A **57** 3518

Photoelectron imaging in pump-probe experiments combining synchrotron and laser radiation

P. O'Keeffe[1], P. Bolognesi[1], R. Richter[2], A. Moise[2], E. Ovcharenko[1,3], L. Pravica[4], R. Sergo[2], L. Stebel[2], G. Cautero[2] and L. Avaldi[1,*]

[1] CNR Istituto di Metodologie Inorganiche e dei Plasmi, Area della Ricerca di Roma 1, CP10,00016 Monterotondo Scalo, Italy
[2] Sincrotrone Trieste, Area Science Park, I-34149 Basovizza (Trieste), Italy
[3] Institute of Electron Physics, Uzghorod, Ukraine
[4] ARC Centre for Antimatter-Matter Studies, Physics Department, The University of Western Australia, Perth 6009, Australia

E-mail: Lorenzo.Avaldi@imip.cnr.it

Abstract. A new photoelectron imaging spectrometer has been used to measure the photoelectron angular distributions in pump-probe experiments where the VUV radiation of the synchrotron is used to prepare an aligned state of the target and the radiation of a Ti:sapphire laser ionizes it. The new set-up and its use to study the photoionization of the He 1s3p ($^1P_1$) and Ne $2p^5$nd ($^1P_1$) states are described.

1. Introduction

Measurements of the *photoelectron angular distributions* (PADs) in the photoionization of atoms and molecules have long been known as a valuable tool to characterise the structure of the continuum and bound electronic states, to provide information on photoionization dynamics and to test theoretical models. Since the early experiments in the 1930s [1], PADs have been used to perform complete photoionization experiments, to observe effects due to site specific emission and alignment of fixed-in-space molecules, and to probe chirality and time-dependent dynamics [2 and references therein]. Renewed interest in the measurement of the PADs has been triggered by the development of new highly efficient imaging detectors [3] and the availability of new sources, such as third generation synchrotron radiation beamlines, fast and intense lasers for multi-photon ionisation and, more recently, by the possibility to combine these two photon sources to tackle processes previously not accessible [4].

In particular, the two-photon photoemission, where the electron ejection proceeds via the excitation and subsequent ionisation of an intermediate state, is an efficient tool for various applications such as *i)* the study of intermediate, neutral excited states [5], *ii)* the measurement of molecular alignment [6], *iii)* the achievement of complete experiments where the phase differences and the dipole matrix elements are determined [7] and *iv)* the investigation of time-resolved photoelectron angular distributions (TRPADs) of intermediate states [8]. However, most of the early studies employed

[*] To whom any correspondence should be addressed.

exclusively laser sources and were confined to targets with suitable intermediate states, matching the available laser wavelengths. We have undertaken a program at the Gasphase photoemission beamline of Elettra to combine the synchrotron radiation with the radiation of a Ti:Sapphire oscillator laser in order to investigate the valence and inner shell states of a broad range of targets via PAD measurements. The use of the synchrotron radiation has several advantages. It is widely and continuously tunable giving access not only to valence but also to core-excited and Rydberg states with energy of tens and even hundreds of eV. Moreover in comparison with the fs pulses of high harmonic generation (HHG) sources it is characterized by a narrower excitation band. The disadvantage is that the duration of the synchrotron bunches limits time-resolved experiments to, at best, a time scale of tens of ps. Within this framework a new velocity map imaging (VMI) spectrometer has been built. Here we present the first results of this new set-up used to investigate the near threshold photoionization of the He 1s3p ($^1P_1$) and Ne 2p3d ($^1P_1$) states.

2. Experiment and data analysis

The experiments have been performed at the Gasphase photoemission beamline at the Elettra storage ring (Trieste, Italy). The lay-out, characteristics and performances of the beamline have been described previously [9] and will be not repeated here. Recently a branch line of the GasPhase beamline has been equipped with a mode-locked Ti:Sapphire oscillator (Tsunami, Spectra Physics) [10] modified to allow operation at a frequency 83.3MHz, which corresponds to exactly 1/6 of the storage ring frequency. By switching off the mode-locking option, the laser can also be used in CW mode. The laser was operated in the picosecond configuration with a measured pulse width of 15 ps, in order to achieve a better temporal overlap with the synchrotron pulses (about 60 ps). In these experiments the polarizations of both radiations have been chosen to be parallel to the detector and perpendicular to the counterpropagating photon beams. The photoelectron yields are very sensitive to the configuration of the electric fields of the exciting and ionizing radiations and Mitsuke et al. [11] showed that the photoionization yield has a maximum in a parallel configuration for the transitions they probed.

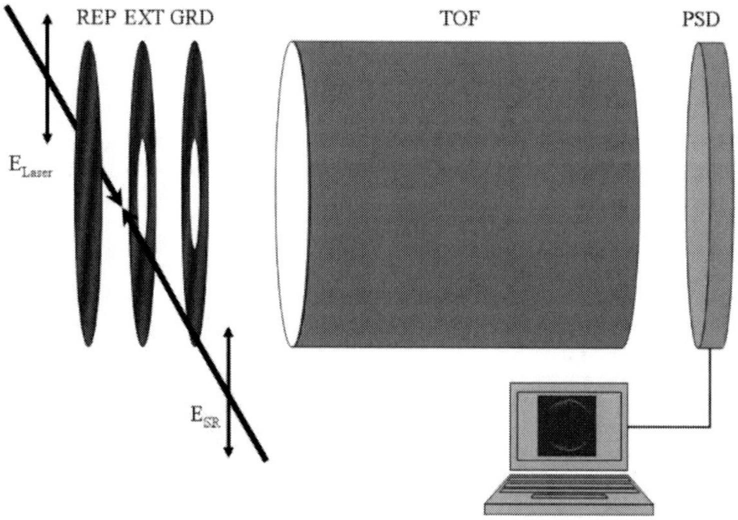

Fig.1: Schematic representation of the VMI spectrometer where REP, EXT and GRD are the repeller, extractor and ground electrodes of the set-up, TOF represents the field-free time of flight region and PSD is the position sensitive detector. The directions of the polarizations of the light sources are also shown.

The present VMI spectrometer design (see figure 1) is based on a classical three element system (20 mm aperture, 15 mm spacing) and a 360 mm time of flight tube, ended by a 95% transparency grid placed in front a set of two MCPs to avoid field penetration effects. The axis of the VMI is perpendicular to the plane containing both the counterpropagating photon beams and their polarization axes. The target gas is let into the experimental chamber via a 0.25 mm inner diameter needle . An x-y-z manipulator allows the precise positioning of the needle near the interaction region. In the operating conditions, the voltages of the extractor and repeller electrodes are both negative, with a fixed ratio of 71% and the needle potential set close to the mean value of the two. A fine tuning of the needle potential was essential to recreate the 'ideal' working conditions of the VMI.
The spatial overlap of the two light beams at the interaction region is very critical for two-color experiments and, because of the nature of the VUV synchrotron radiation, it had to be optimized and monitored under vacuum. To this purpose a specifically designed alignment "tool" with a set of 1.5, 1.0 and 0.5 mm holes has been mounted on the x-y-z manipulator of the gas inlet and moved to the interaction region to check the overlap of the two beams. The spot size of the laser beam has not been measured. The beam was focused to the center of the chamber using a 50 cm focal length lens. Thus, the size of the focus can be estimated to have a diameter of 0.25 mm, which is of the same order of magnitude as the synchrotron beam size. The estimated power density of the laser is 1 MW/cm^2 (when mode-locked).
The spectrometer is equipped with a position sensitive detector [12] developed by the Instrumentation and Detector Laboratory at Elettra. The UHV section of the detector consists of two 50 mm MCP mounted in chevron configuration, a high voltage decoupling system based on a resistive sheet and a cross delay line detector. The spatial resolution is about 100 μm in one direction and 150 μm in the other. The X,Y start and stop signals from the detectors are immediately amplified at the UHV flange. The amplified pulses are sent to constant fraction discriminators and then fed to a time-to-digital converter (TDC). An ACAM TDC-GPX has been chosen because, according to the selected operation mode, up to 10 ps time resolution with a broad time window and high count rate (deadtime of the TDC is less than 10 nsec when operating in multihit mode) can be achieved. Two TDCs, one for each detector dimension, are used. The TDCs are controlled by a mother board hosting a field programmable gate array (FPGA), which handles the flux of events, controls their logical consistency, correlates them with "time stamps" and transfers the digitized data. The motherboard is connected via an external SCSI cable to a PCI Interface board inside a PC. Finally a Labview code manages the data acquisition, analysis and on-line graphical presentation.
The chamber is lined with a 2mm thick μ-metal shield. This allows the measurement of the angular distribution of photoelectrons with kinetic energy as low as 13 meV with a typical energy resolution of 5%. Under normal working conditions, a gas pressure of about $3x10^{-7}$ mbar was used, with synchrotron photon beam current of about 10^{11} photons/s. This setting was used to avoid saturation of the MCP, due to the high efficiency of the 4π collection of the spectrometer. Typical acquisition times of an image were of the order of 5 minutes for one photon and 40 minutes for singly resonant two-photon ionisation processes.
The VMI spectrometer works by projecting the 3D distribution of electrons released in the ionization process onto the detector. It is then possible to recover the PAD and kinetic energy distribution (KED) by performing an inverse Abel transformation or other inversion methods such as the onion-peeling [13], Basex [14], or pBasex [15] methods. The data analysis in this work was performed using the pBasex method introduced by Garcia *et al.* [15] in which the original 3D image is reconstructed as a linear combination of basis set functions adapted to the cylindrical symmetry of the photoionization process. Furthermore, we have ported the code of Garcia *et al.* [15] into a new Windows type graphical user interface which has extensive image manipulation features.
The reported anisotropy parameters are determined using a weighted average of all points contained within the full width half maximum of the peak in the kinetic energy distribution (as suggested in ref. [15]). The errors are then determined based on the spread of the anisotropy parameters extracted by varying the centre point of the image and other parameters related to the image processing.

3. Results

In one-photon ionisation of randomly oriented targets by fully linearly polarized radiation the photoelectron angular distribution is represented by the double differential cross section

$$\frac{d^2\sigma(E,\theta)}{dEd\Omega} = \frac{\sigma_0}{4\pi}[1+\beta P_2(\cos\theta)] \qquad (1)$$

where σ_0 is the total photoionization cross section, θ the angle of the photoelectron with respect to the polarization direction of the incident radiation, $P_2(\cos\theta)$ is the second order Legendre polynomial and β the asymmetry parameter. The β parameter holds information on the photoionization dynamics, because it depends on the radial matrix elements, $R_{l\pm1}$, and their relative phase, $\delta_{l-1}-\delta_{l+1}$, where $l\pm1$ are the final state partial waves of the free photoelectron in the continuum. The interference between the $l\pm1$ partial waves determines the variation of β with energy in its allowed range (-1,2). However, despite the fact that β contains the full information on the photoionization matrix elements and their relative phase, its determination alone does not allow this information to be extracted.

In two-photon ionisation when both photon beams have parallel polarization, the angular distributions are of the form [5]:

$$\frac{d^2\sigma(E,\theta)}{dEd\Omega} = \frac{\sigma_0}{4\pi}[1+\beta_2 P_2(\cos\theta)+\beta_4 P_4(\cos\theta)] \qquad (2)$$

where both the β_2 and β_4 anisotropy parameters are combinations of $R_{l\pm1}$, and $\delta_{l-1}-\delta_{l+1}$. Thus an experimental determination of β_2 and β_4 allows the extraction of at least two (even though not all) independent quantities governing the photoionization, i.e. the ratio $\chi=R_{l+1}(j)/R_{l-1}(j)$ of the matrix elements and their relative phase.

3.1 The photoionization of He 1s3p($^1P_1$) state

In this measurement the He $1s^2$ ($^1S_0$)$\rightarrow$1s3p ($^1P_1$) transition at 23.09 eV was resonantly excited by synchrotron radiation and then ionised by a synchronised laser pulse of 750 nm, producing a photoelectron of 160 meV kinetic energy. The long lifetime (1.7 ns[16]) of the state guarantees a large population of the excited state during the probing with the IR laser. The raw and reconstructed images of the two-photon photoelectron angular distribution are shown on the right side of fig.2, panels b and d respectively. The comparison to the 'equivalent' one-photon distributions, i.e. at the same photoelectron kinetic energy, (fig. 2a and c) shows clear differences between the two cases, with a much sharper angular distribution in the two-photon case. Indeed, in the one-photon ionization the photoelectron is constrained to a *p* partial wave resulting in a dipolar, β=2, angular distribution. In the two-photon ionization case, the angular distribution is described by the β_2 and β_4 anisotropy parameters (see formula 2), each one resulting from the energy dependent interference of both the *s* and *d* photoelectron partial waves providing a more complex picture. The extracted anisotropy parameter for the one-photon distribution is $\beta=1.98\pm0.02$. This value confirms the reliability of the VMI set-up, the inversion and analysis procedures and gives a hint of the accuracy of the measured

Table 1. The anisotropy parameters for the photoionization of He 1s3p ($^1P_1$) intermediate state.

Photoelectron kinetic energy (eV)	β_2	β_4
0.04	2.88±0.06[a]	1.93±0.07[a]
0.16	2.86±0.07[b]	2.17±0.13[b]
1.6	2.81±0.05[a]	1.88±0.13[a]
3.14	2.55±0.17[a]	1.78±0.2[a]

[a] values reported in ref.[17]; [b] present work

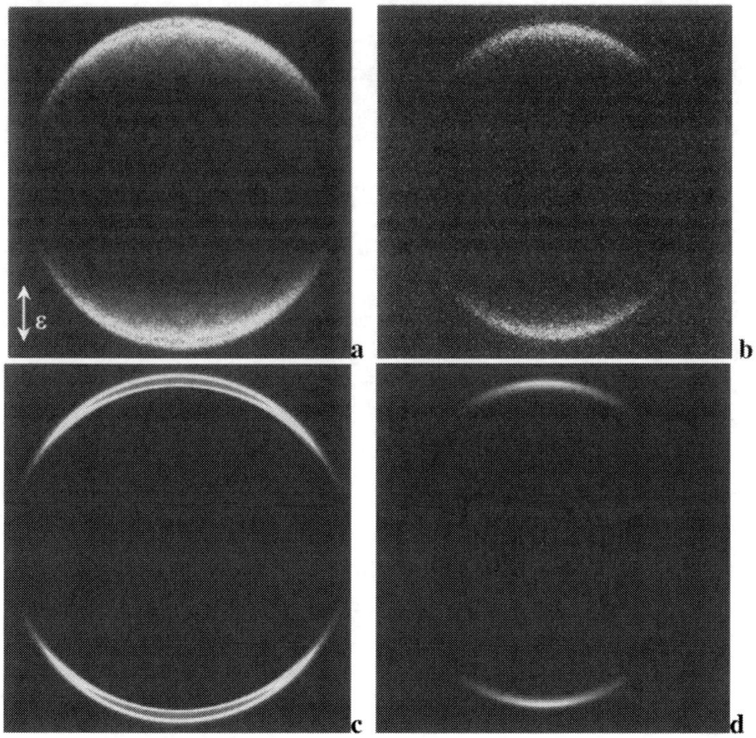

Fig.2: Left hand side: the raw photoelectron image (a) and its pBasex reconstruction (c) following He one-photon ionization. Right hand side: the raw photoelectron image (b) and its pBasex reconstruction (d) following two-photon ionization of He via the 1s3p intermediate state. In both cases the kinetic energy of the photoelectron is 160 meV.

anisotropy parameter. The extracted values of the anisotropy parameters for the two-photon distribution are β_2=2.86±0.07 and β_4=2.17±0.13.

Recently Haber et al. [17] measured the two-photon ionization of He via the same intermediate state using high-order harmonics for the excitation of the target and a laser at three different wavelengths (800, 400 and 267 nm) for the final ionization. Their results are reported in table 1 and figure 3, where they are compared with the present determination and the calculated values [17,18]. Good agreement between the two set of measurements is observed in the β_2 case, while a slightly higher value of β_4 has been measured in the present work. In two-photon processes the values of the anisotropy parameters are not limited to the (-1,2) range and indeed in the range studied the β_2 value is always larger than 2. The calculations show an increase of β_2 for photoelectron energies smaller than 0.3 eV and then a monotonic decrease. A similar behaviour, but for values smaller than 2 is observed for β_4.

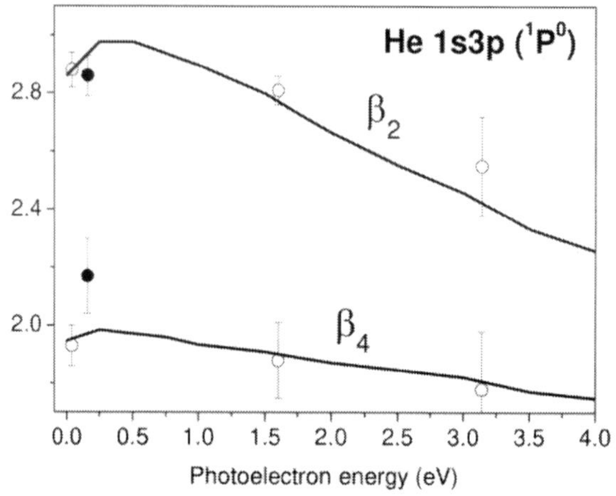

Fig.3: The β_2 and β_4 anisotropy parameters for the photoionization of the He 1s3p ($^1P_1$) excited state. The present values (full dots) are compared with previous experimental results (open circles) and calculations [17].

3.2 The photoionization of Ne 2p3d $[3/2]_1$ in the threshold region

In the Ne case the synchrotron radiation was set to 20.04 eV, in order to excite the Ne $2p^6$ ($^1S_0$)$\rightarrow$$2p^5$($^2P^0_{1/2}$)3d$[3/2]_1$ transition, which is characterized by a quite large oscillator strength (1.86($\pm$0.09)x10^{-2} [19]), and then several laser wavelengths between 775 and 805 nm have been chosen to ionize the excited state. Thus all the measurements are within 100 meV from the Ne^+ $2p^{-1}$ ($^2P_{3/2}$) ionization threshold, ie. below the Ne^+ $2p^{-1}$ ($^2P_{1/2}$) ionization threshold. In this way only one continuum is open and the anisotropy parameters are not affected by the multiple interactions between the Rydberg electron and the other electrons, which can lead to a change of the angular momentum of the ion core. The measured β_2 and β_4 values are reported in Table 2 and figure 4.

Table 2. The β_2 and β_4 anisotropy parameters for the photoionization of Ne 2p3d $[3/2]_1$ intermediate state.

Photoelectron kinetic energy (meV)	β_2	β_4
20	0.84 ± 0.06	0.42 ± 0.02
50	0.87 ± 0.03	0.40 ± 0.02
70	0.93 ± 0.08	0.39 ± 0.04

To our best knowledge the only calculation of the anisotropy parameters for the photoionization of a Ne excited state is the one due to Chang et al [20] for Ne $2p^5$3p while no data exists for the Ne$2p^5$nd states. The values of the anisotropy parameters β_2 and β_4 are smaller than those observed and calculated for the excited state of He because here different partial waves (*p* and *f*) are involved. The β_2 values increase for photon energies away from threshold, while β_4 stays almost constant in the region investigated.

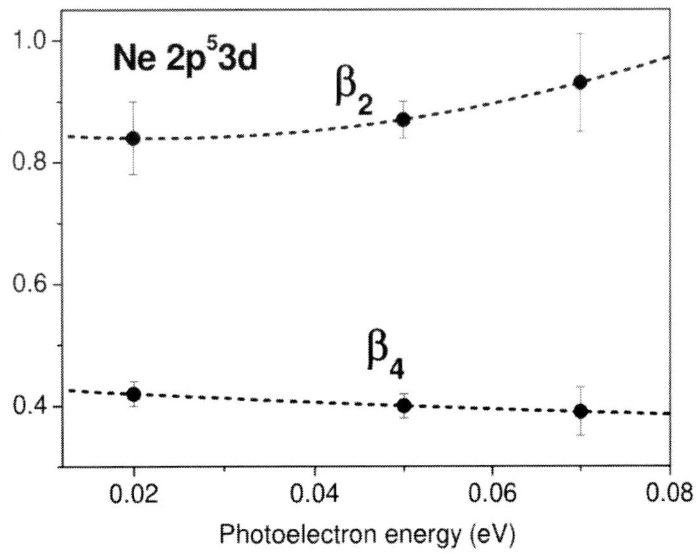

Fig.4: The β_2 and β_4 anisotropy parameters for the photoionization of the Ne $2p^5 3d$ $[3/2]_1$ excited state. The dashed lines through the experimental points serve only to guide the eyes

4. Conclusions

A novel experimental set-up using a VMI spectrometer in conjunction with a VUV synchrotron radiation + IR laser source has been used to measure the photoelectron angular distribution in two-photon - two-colour ionization experiments. The photoelectron angular distribution from the excited Ne $2p^5(^2P^0_{3/2})3d$ state has been studied in the first 100 meV above threshold, a region rarely studied [*21*] in previous photoionization studies with synchrotron radiation. The independent-particle models, often used to describe photoionization of atoms [*22*] are expected to break down near threshold and in the case of non negligible many-electron interaction. Thus the present measurements on Ne may be used to extend the successful test of theoretical predictions done in He [17] to a more challenging situation.

The high efficiency and the capability of handling very low energy electrons make the VMI technique a suitable tool to investigate the dynamics of molecular dissociation via time resolved photoelectron angular distribution and pump-probe experiments where ps time delays are introduced between the synchrotron excitation of the target and its probe with the IR laser.

5. Acknowledgments

LP acknowledges the support of the CNR-Short Term Mobility Program and the Australian Academy of Science for an Australia-Italy Award

References

[1] Lawrence EO and Chaffee MA 1930 *Phys. Rev.* **36** 1099
[2] Reid KL 2003 *Annu. Rev. Phys. Chem.* **54** 397
[3] Eppink AT and Parker JB 1997 *Rev. Sci. Instrum.* **68** 3477
[4] Meyer M *et al* 1987 *Phys. Rev. Lett.* **59** 2963

[5] Hansen JC *et al* 1980 *Phys Rev. A* **21** 222; Eldestein S *et al* 1974 *Phys. Rev.* A **9** 2459; Kaminski H *et al* 1980 *Phys. Rev. Lett.* **45** 1161
[6] Appling JR, White MG, Orlando TM and Anderson SL 1986 *J. Chem. Phys.* **85** 6803
[7] Strand MP, Hansen J, Chien R L and Berry RS 1978 *Chem. Phys. Lett.* **9** 205
[8] Davies JA, Continetti RE, Chandler DW and Hayden CC 2000 *Phys. Rev. Lett.* **84** 5983
[9] Blyth RR *et al* 1999 J. *Electron Spectrosc. Relat. Phenom.* **101-103** 959
[10] Moise A, Alagia M, Banchi L, Ferianis M, Prince KC and Richter R 2008 *Nucl. Instr. Meth. A* **588** 502
[11] Mitsuke K, Hikosaka Y and Iwasaki K 2000 *J. Phys. B:At. Mol. Opt. Phys.* **33** 391
[12] Cautero G, Sergo R, Stebel L, Lacovig P, Pittana P, Predonzani M and Carrato S 2008 *Nucl. Instr. Meth. A* **595** 447
[13] Vrakking MJJ 2001 *Rev. Sci. Instrum.* **72** 4084
[14] Dribinski V, Ossadtchi A, Mandelstam VA, and Reisler H 2002 *Rev. Sci. Instrum.* **73** 2634
[15] Garcia GA, Nahon L and Powis I 2004 *Rev. Sci. Instrum.* **75** 4989 *(2004)*
[16] Johansson A *et al* 2003 *Eur. Phys. J. D* **22** 3
[17] Haber LH, Doughty B and Leone SR 2009 *Phys. Rev.* A**79** 031401(R)
[18] Chang TN and Fang TK 1995 *Phys. Rev.* A **52** 2638
[19] Chan WF, Cooper G, Guo X and Brion CE 1992 *Phys. Rev.* **A** 45 1420
[20] Chang TN and Kim YS 1982 *Phys. Rev.* A **26** 2728
[21] Czasch A *et al* 2005 *Phys. Rev. Lett.* **95** 243003
[22] Cooper J and Zare RN 1968 *J. Chem. Phys.* **48** 942

International Workshop on Electronic Spectroscopy for Gas-phase Molecules and Solid Surfaces IOP Publishing
Journal of Physics: Conference Series **235** (2010) 012007 doi:10.1088/1742-6596/235/1/012007

The valence shell excitations for some noble atoms studied by fast electron impact

Lin-Fan Zhu
Hefei National Laboratory for Physical Sciences at Microscale, Department of Modern Physics, University of Science and Technology of China, Hefei, Anhui, 230026, P. R. China
E-mail: lfzhu@ustc.edu.cn

Abstract.
The differential cross sections and apparent generalized oscillator strengths for the valence-shell excitations of the noble atoms of He, Ne, Ar, and Kr are reviewed. The asymptotical behaviors of the apparent generalized oscillator strengths show that the first Born approximation is more easily reached for the dipole-allowed transition than the dipole-forbidden one of $^1S_0 \rightarrow ^1 S_0$ type, in which the final state has the same term symbol of the initial state. Furthermore, the experiments reveal that the generalized oscillator strengths for the transitions, which are formed by the same electronic configuration through the intrachannel interaction such as $(n-1)p^5ns[3/2]_1$ and $(n-1)p^5ns'[1/2]_1$ of Ne (n=3), Ar (n=4), and Kr (n=5), are nearly parallel. This phenomenon is elucidated by the fact that the LS coupling singlet nature of the ground state selects the LS coupling singlet component from the excited wave function, i.e., the contribution from the singlet component in the excitation is dominant in the generalized oscillator strength.

1. Introduction

The cross sections of electron impact excitation are important in the diverse fields such as radiation physics, plasma physics, atmospheric physics, and astrophysics, and there are ever-increasing needs for such cross-section data. For fast electron impact, the first Born approximation (FBA) is applicable if the incident electron energy is high enough but still nonrelativistic. The differential cross section (DCS) can be factorized into two factors: one dealing with the incident particle only, which is a trivial factor; the other dealing with the target only, which is the so-called generalized oscillator strength (GOS) [1, 2]. The GOS is a fundamental quantity for atoms and molecules since it relates to various physical constants and is very sensitive to the structure of the target. Experimentally, the GOS can be obtained directly from the DCS measured by the electron scattering experiment when the FBA condition is satisfied. However, for the slow electron impact, the FBA condition is not satisfied, then the GOS can not be derived from the experimental DCS and the influence of the incident electron should be considered. Therefore, the investigations of the GOS, as well as the condition of the FBA, is an important content for atomic and molecular physics [2].

In this paper the DCSs and apparent GOSs for the valence-shell excitations of He, Ne, Ar, and Kr measured at the higher impact energies ($\geq$ 100eV) are reviewed. For helium, it is the simplest mutli-electron system, and its experimental DCSs and GOSs can test the theoretical method and the theoretical code stringently. Therefore, its DCSs and GOSs were extensively investigated both experimentally [3, 4, 5, 6, 7, 8, 9, 10, 11, 12, 13] and theoretically [14, 15, 16]. Among them, Trajmar and his collaborators [3, 4] have measured the DCSs of electron-impact excitation of helium at the incident electron energies lower than 100 eV, while Dillon and Lassettre [5, 6] and Suzuki and his coworkers [7, 8, 9] measured the DCSs of the 2^1S_0, 2^1P_1, and 2^3S_1 states at moderate kinetic energies of 100-700 eV. In our group, we measured the DCSs and GOSs of the excitations of helium at much higher impact energies of 1500 eV and 2500 eV [10, 11, 12, 13]. Recently, the GOSs of the 2^1S_0 and 2^1P_1 states of helium were measured by me and my collaborators using the inelastic X-ray scattering (IXS) [17]. The theoretical calculations were carried out by the R-matrix theory [14], the first Born approximation [15], and the convergent close-coupling theory [16]. Furthermore, it should be mentioned that the DCSs and GOSs for the excitations of helium have been measured and calculated with a high accuracy [12, 14, 15, 18], so the reliable DCSs and GOSs of He are often used as the standard data to normalize the DCSs and GOSs for other atoms and molecules [18, 19].

For the noble atoms of Ne, Ar and Kr, their DCSs and GOSs were measured by Refs. [9, 19, 20, 21, 22, 23, 24, 25, 26, 27, 28, 29, 30]. Among them, Khakoo and his collaborators [20, 21] reported the DCSs of electron-impact excitation of Ne and Ar at the incident electron energies lower than 100 eV, while Suzuki and his coworkers [9, 22, 23, 24] measured the DCSs of the valence-shell excitations of Ne, Ar and Kr at moderate kinetic energies of 100-500 eV. In our group, we measured the DCSs and GOSs for the valence-shell excitations of Ne, Ar and Kr at the high impact energies of 1500 eV and 2500 eV [19, 25, 26, 27, 29, 30]. As for the theoretical studies, the DCSs and GOSs were calculated by the random phase approximation with exchange (RPAE) methods [31, 32, 33]. Since the excited energy level structures of Ne, Ar and Kr are well described by the intermediate coupling, while their ground states are closed-shell $^1S_0$ states of LS coupling, these noble atoms are the typical systems to study the valence-shell excitation mechanism and dynamics relating to two different coupling scheme, which were explored by our group recently [19, 26, 34].

In this paper, the DCSs and GOSs for electron impact excitation of the valence-shell states of He, Ne, Ar and Kr at the moderate and high impact energis are reviewed. In the following sections, the theoretical background will be given in Sec. 2. Then, the apparent GOSs of helium will be reviewed in Sec. 3, and the momentum transfer dependence behaviors for Ne, Ar and Kr will be discussed in Sec. 4. Finally some conclusions will be included in Sec. 5.

2. Theoretical background

When the incident electron is sufficiently fast, i.e., the first Born approximation is valid and the incident and scattered electrons can be treated as plan waves, the differential cross section $d\sigma_n^e/d\Omega$ of a state n of an atom or molecule excited by the fast electron impact can be described by [1, 2] (in atomic unit):

$$\frac{d\sigma_n^e}{d\Omega} = \frac{2}{E_n}\frac{p_a}{p_0}\frac{1}{K^2}f(E_n, K^2), \tag{1}$$

where $f(E_n, K^2)$ is the generalized oscillator strength defined as

$$f(E_n, K^2) = \frac{2E_n}{K^2}\left|\langle\Psi_n|\sum_{j=1}^{N} e^{i\vec{K}\cdot\vec{r_j}}|\Psi_0\rangle\right|^2. \tag{2}$$

Here E_n and K are the excitation energy and the momentum transfer, while p_0 and p_a stand for the incident and scattered electron momenta, respectively. Ψ_0 and Ψ_n are N electrons wave functions for the initial and final states, respectively. $\mathbf{r_j}$ is the position vector of the jth electron. It can be seen that for sufficiently fast electron impact, the influence of the incident particle upon an atom or molecule can be regarded as a sudden and small external perturbation, and the exchange effect is negligibly small [1, 2]. Differently, for the slow electron impact, the collision problem is considered as a combined system of the incident particle and the target, in which both of them lose their mechanical individuality. Even so, the DCSs measured at the intermediate impact energies can be converted into the oscillator strengths, i.e., the so-called apparent generalized oscillator strength (AGOS) $f^A(E_0, E_n, K^2)$, with a similar form as Eq. (1):

$$f^A(E_0, E_n, K^2) = \frac{E_n}{2}\frac{p_0}{p_a}K^2\frac{d\sigma_n^e}{d\Omega}. \tag{3}$$

With the increasing of the incident electron energies, the measured apparent GOSs should approach to the same curve, which means that the FBA is valid and the measured apparent GOSs approach to GOS:

$$f^A(E_0, E_n, K^2)|_{E_0\to\infty} = f(E_n, K^2). \tag{4}$$

Then the measured GOS of $f(E_n, K^2)$, which is free of the influence of the incident electron, reveals the momentum distribution character related to the wave functions of the initial and final states ($|\Psi_0\rangle$ and $|\Psi_n\rangle$), and it provides the benchmark to testify the theoretically calculated wave functions [2, 14].

Another experimental method to explore the character of the wave functions of the initial and final states is the Compton excitation, in which the DCSs of the excited states are measured through the inelastic X-ray scattering or the so-called nonresonant X-ray Raman scattering [35, 36, 37]. The differential cross section $d\sigma_n^r/d\Omega$ of the Compton excitation can be described by:

$$\frac{d\sigma_n^r}{d\Omega} = r_0^2\frac{\omega_f}{\omega_i}|\vec{\epsilon_i}\cdot\vec{\epsilon_f^*}|^2 S(\vec{q},\omega), \tag{5}$$

where $S(\vec{q},\omega)$ is the dynamical structure factor defined as

$$S(\vec{q},\omega) = \left|\langle\Psi_n|\sum_{j=1}^{N} e^{i\vec{q}\cdot\vec{r_j}}|\Psi_0\rangle\right|^2. \tag{6}$$

Here r_0 is the classic electron radius, while $\hbar\omega_i$ and $\hbar\omega_f$ are the energies of incident and scattered X-ray, respectively. $\hbar\omega = \hbar\omega_i - \hbar\omega_f$ is the energy loss, and $\vec{q}$ is the momentum transfer (which is called $\vec{K}$ in the electron impact experiment). $\vec{\epsilon_i}$ and $\vec{\epsilon_f}$ stand for the polarization directions for the incident and scattered photons.

Considering $E_n = \hbar\omega$ and $\vec{K} = \vec{q}$, the GOS defined in Eq. (2) can be rewritten as (in atomic unit):

$$f(\omega, q^2) = \frac{2\omega}{q^2}S(\vec{q},\omega) = \frac{1}{r_0^2}\frac{\omega_i}{\omega_f}\frac{2\omega}{q^2}\frac{1}{|\vec{\epsilon_i}\cdot\vec{\epsilon_f^*}|^2}\frac{d\sigma_n^r}{d\Omega}. \tag{7}$$

It can be seen from Eqs. (1-2) and (5-7) that there is an intrinsic connection between the fast electron scattering and the inelastic X-ray scattering. Therefore, we can convert the DCS measured by IXS into GOS, which provide a new and independent method to determine the GOS, and the GOSs measured by these two methods can provide independent cross-check. The merit of the IXS method is that the FBA is always valid, which avoids the tedious procedures to judge the validity of FBA on a trial-and-error basis in the electron impact method. The demerit of the IXS method is the low cross section of the Compton excitation.

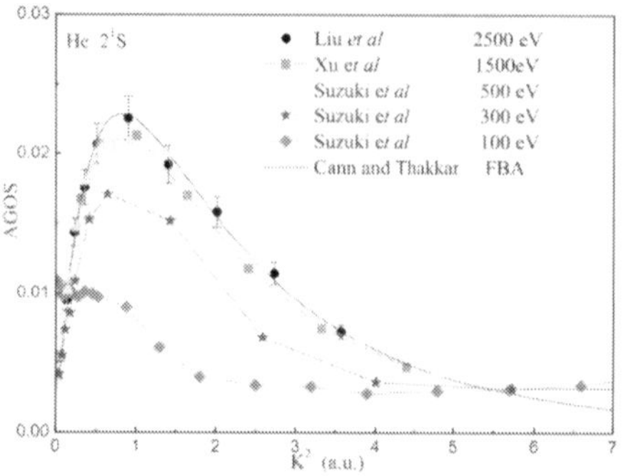

Figure 1. The apparent GOSs for the 2^1S_0 excitation of He. The experimental results are taken from Suzuki *et al* [7], Xu *et al* [10] and Liu *et al* [12], while the theoretical results are taken from Cann and Thakkar [15].

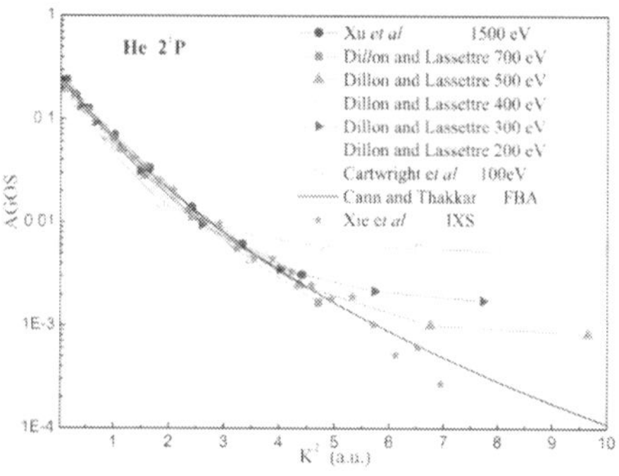

Figure 2. The apparent GOSs for the 2^1P_1 excitation of He. The experimental results are taken from Cartwright *et al* [3], Dillon and Lassettre [5], Xu *et al* [10] and Xie *et al* [17], while the theoretical results are taken from Cann and Thakkar [15].

3. Generalized oscillator strengths for He

The DCSs for the excitations of 2^1S_0 and 2^1P_1 of He were extensively investigated both experimentally [3, 5, 7, 9, 10, 11, 12] and theoretically [14, 15]. Herein only the recent apparent GOSs of 2^1S_0 and 2^1P_1 measured at the incident electron energies from 100 eV to 2500 eV are shown in Fig. 1 and 2 along with the theoretical GOSs calculated within the FBA [15], which are in a good agreement with other calculations [14]. It can be seen from Fig. 1 and 2 that the apparent GOSs for the both excitations approach to the GOSs with the increasing of the impact energies, which is the typical behaviors of apparent GOS. It can also be seen that for the dipole-allowed transition of 2^1P_1 the apparent GOSs in the low K^2 region are very close to the GOSs even for the impact energy as low as 200 eV, while for the dipole-forbidden transition of 2^1S_0 the apparent GOSs are systematically lower than the GOSs even for the impact energy as high as 1500 eV. These results mean that the FBA is more easily reached for the dipole-allowed transition than the dipole-forbidden one of $^1S_0 \rightarrow ^1S_0$ type. The above phenomenon can be elucidated by that the higher order Born amplitudes give the main contribution to the apparent GOS for the transition in which the terms of the initial and final states are the same, as the case of the $^1S_0 \rightarrow ^1S_0$ transition [7, 9].

Fig. 2 also shows the GOSs measured by IXS [17]. Because the FBA is always valid for IXS method, the results measured by IXS can serve for the standard GOS data. The situation is true because the GOSs measured by IXS show a very good agreement with the theoretical calculations [14, 15] and the results measured by a high energy electron impact [10]. Therefore the work of Xie *et al* [17] shows that the IXS is a new and independent method to determine the GOS at a third generation synchrotron. The independent cross-check shows that the theoretical calculations [14, 15] are reliable and the FBA holds for the measured K^2 region of 0-4.4 a.u. of the 2^1P_1 transition at an impact energy of 1500 eV.

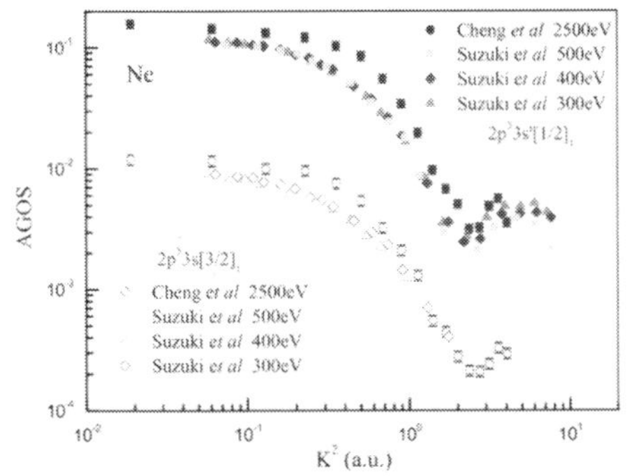

Figure 3. The apparent GOSs for the $2p^53s[3/2]_1$ and $2p^53s'[1/2]_1$ excitations of Ne. The experimental results were taken from Cheng *et al* [19] and Suzuki *et al* [22].

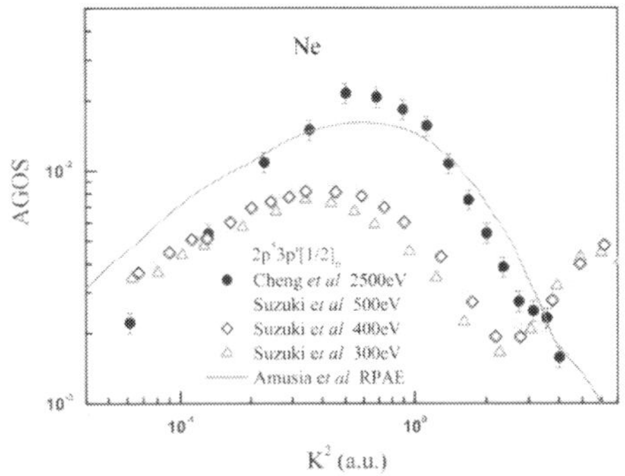

Figure 4. The apparent GOSs for the $2p^53p'[1/2]_0$ excitation of Ne. The experimental and theoretical results were taken from Cheng *et al* [19], Suzuki *et al* [22] and Amusia *et al* [31].

4. Momentum transfer dependence behaviors for Ne, Ar and Kr

4.1. Generalized oscillator strengths of Ne, Ar and Kr

The apparent GOSs for the valence-shell excitations of Ne were measured by Refs. [19, 20, 22] and calculated by Refs. [31, 32, 33]. Fig. 3 shows the measured apparent GOSs for the excitations of $2p^53s[3/2]_1$ and $2p^53s'[1/2]_1$ of Ne [19, 22]. It can be seen that the apparent GOSs measured at a much higher impact energy of 2500 eV [19] are systemically higher than the ones of Suzuki *et al* [22] measured at the impact energies of 300, 400 and 500 eV. The discrepancies may be due to the different normalization procedures: Suzuki *et al* [22] normalized their data on the elastic DCSs of Ne while Cheng *et al* [19] normalized their data on the absolute GOSs of 2^1P_1 of He calculated by Cann and Thakkar [15], which are in good agreement of the experimental results as shown in Section 3. Furthermore, the theoretical calculation by the random phase approximation with exchange (RPAE) methods [32, 33] supports the results measured at 2500 eV. Whereas, if we exclude the possible normalization problems, the FBA is valid at the lower K^2 region when the impact energy is larger than 300 eV since the apparent GOSs at the impact energies of 300, 400 and 500 eV measured by Suzuki *et al* [22] are in the same curve. The apparent GOSs for the dipole-forbidden excitation of $3p'[1/2]_0$ of Ne are shown in Fig. 4, and it can be seen that there are apparent discrepancies including the shapes and the extremal positions between the apparent GOSs measured at the different impact energies, which means that the FBA is more difficult to reach for the dipole-forbidden excitation of $3p'[1/2]_0$ than the dipole-allowed ones of $2p^53s[3/2]_1$ and $2p^53s'[1/2]_1$. This phenomenon is similar with the one of He discussed above, since the $2p^53p'[1/2]_0$ state in Ne is well represented by the LS term $^1S_0$, which is the same as the term $^1S_0$ of the ground state of $2p^6$ [9, 19]. The RPAE calculation [31] is in general agreement with the experimental one measured at 2500 eV. Therefore, we have the reason to believe that the FBA holds at an impact energy of 2500 eV for the valence excitations of Ne.

The apparent GOSs for the valence-shell excitations of Ar were measured by Refs. [21, 23, 25, 26, 27, 28] and calculated by Refs. [31, 32]. Fig. 5 shows the measured apparent GOSs for the excitations of $3p^54s[3/2]_1$ and $3p^54s'[1/2]_1$ of Ar [23, 25, 26]. The apparent GOSs measured at different incident electron energies are in general agreement, while the small

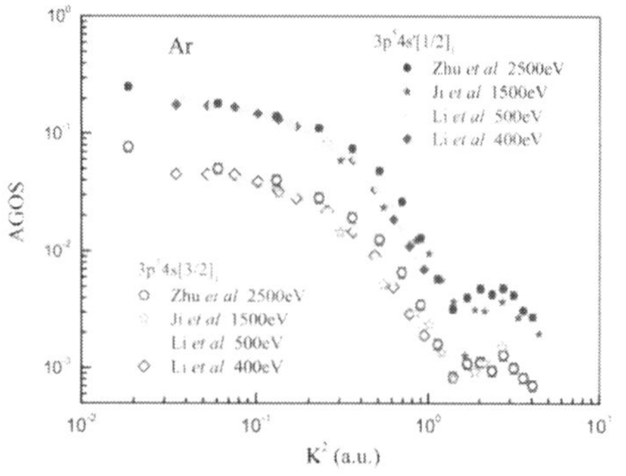

Figure 5. The apparent GOSs for the $3p^54s[3/2]_1$ and $3p^54s'[1/2]_1$ excitations of Ar. The experimental results were taken from Zhu *et al* [26], Ji *et al* [25] and Li *et al* [23].

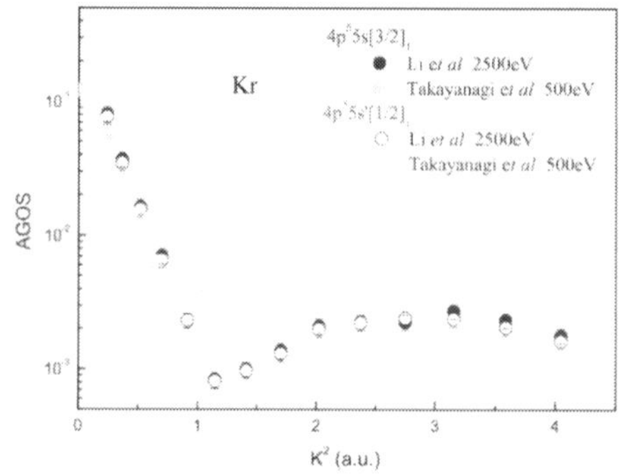

Figure 6. The apparent GOSs for the $4p^55s[3/2]_1$ and $4p^55s'[1/2]_1$ excitations of Kr. The experimental results were taken from Li *et al* [29] and Takayanagi *et al* [24].

differences for the large K^2 region may be attributed to the fact that the FBA does not hold in this K^2 region. The total apparent GOSs of $3p^54s[3/2]_1$ and $3p^54s'[1/2]_1$ measured by Zhu *et al* [26] show a good agreement with the ones of Fan and Leung [28], in which the different normalization procedure of the TRK sum rule is used. The shape and extremal positions of the GOSs calculated by RPAE [32] are in a good agreement with the experimental results of Zhu *et al*, so we have the confidence to believe that the FBA is satisfied at an impact energy of 2500 eV in the measured K^2 region.

The apparent GOSs for the valence-shell excitations of Kr were measured by Refs. [9, 24, 29] and calculated with RPAE by Refs. [31, 32]. Fig. 6 shows the measured apparent GOSs for the excitations of $4p^55s[3/2]_1$ and $4p^55s'[1/2]_1$ of Kr [29, 24]. Although the apparent GOSs measured at an impact energy of 2500 eV [29] are apparently different from the ones determined at 500 eV [24], the former is more reliable and satisfies the FBA since it is measured at a much higher impact energy and perfectly matchs the theoretical calculation based on RPAE [29, 32, 33].

4.2. Generalized oscillator strength ratios of Ne, Ar and Kr

It can be seen from Fig. 3, 5 and 6 that the GOS curves of $(n-1)p^5ns[3/2]_1$ and $(n-1)p^5ns'[1/2]_1$ of Ne (n=3), Ar (n=4), and Kr (n=5) measured at 2500 eV are nearly parallel. Considering the simpleness of the physical picture of the FBA for the fast electron impact excitation, this common phenomenon for Ne, Ar and Kr is strange because the wave functions for the two excited states of every atom are different although its wave function for the ground state is same. In order to elucidate the phenomenon, the wave functions of the relevant states of Ne, Ar, and Kr were calculated using Cowan code [19, 26, 34]. Then the intermediate coupling wave functions for the excited states can be expanded by the linear combination of the basis of the *LS* coupling wave function:

$$\Psi_{ns[3/2]_1} = \alpha_n \Psi_{ns^3P_1} + \beta_n \Psi_{ns^1P_1} \quad (8)$$

$$\Psi_{ns'[1/2]_1} = -\beta_n \Psi_{ns^3P_1} + \alpha_n \Psi_{ns^1P_1}. \quad (9)$$

Here $\Psi_{ns[3/2]_1}$ and $\Psi_{ns'[1/2]_1}$ are the wave functions of the excited states of $ns[3/2]_1$ and $ns'[1/2]_1$ in the intermediate coupling scheme, and $\Psi_{ns^3P_1}$ and $\Psi_{ns^1P_1}$ are the wave function basis in the *LS* coupling scheme while α_n and β_n are the corresponding intermediate coupling coefficients.

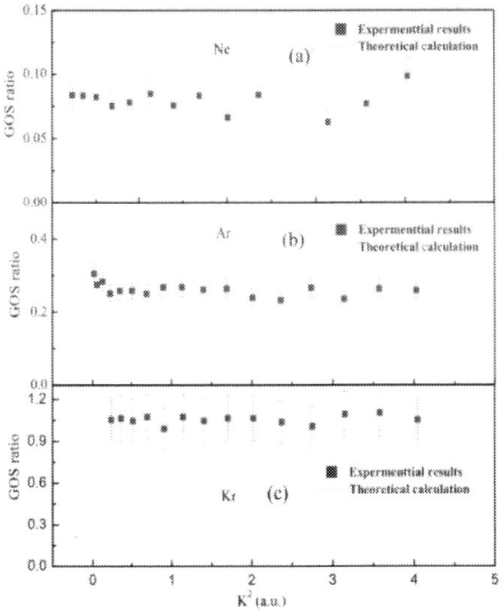

Figure 7. The experimental and calculated GOS ratios between the excitations of $(n-1)p^5ns[3/2]_1$ and $(n-1)p^5ns'[1/2]_1$ of (a) Ne (n=3) [19]; (b) Ar (n=4) [26]; and (c) Kr (n=5) [34].

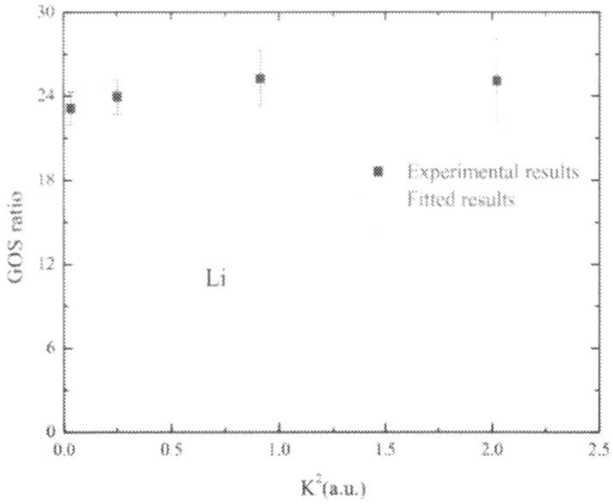

Figure 8. The experimental GOS ratios between the inner excitations of $1s2s(^3S_1)2p\ ^2P$ and $1s2s(^1S_0)2p\ ^2P$ of lithium [38].

Since the spin-forbidden transition is negligibly small for fast electron impact [2] and the initial ground state is singlet, the contribution of the triplet component in the excitation to GOS is negligibly small, i.e., $<\Psi_{ns^3P_1}|\sum_{j=1}^{N} e^{i\vec{K}\cdot\vec{r_j}}|\Psi_{(n-1)p^6\,^1S_0}>=0$. Since the momentum transfers at the same scattering angle are nearly equal for the two transitions, the GOS ratio δ can be obtained according to the definition of GOS:

$$\begin{aligned}\delta &= \frac{f_{ns[3/2]_1}}{f_{ns'[1/2]_1}} = \frac{E_1}{E_2}\frac{|<\beta_n\Psi_{ns^1P_1}|\sum_{j=1}^{N} e^{i\vec{K}\cdot\vec{r_j}}|\Psi_{(n-1)p^6\,^1S_0}>|^2}{|<\alpha_n\Psi_{ns^1P_1}|\sum_{j=1}^{N} e^{i\vec{K}\cdot\vec{r_j}}|\Psi_{(n-1)p^6\,^1S_0}>|^2} \\ &= \frac{E_1}{E_2}\frac{\beta_n^2}{\alpha_n^2},\end{aligned} \tag{10}$$

where $f_{ns[3/2]_1}$, E_1 and p_1 are the GOS, excitation energy and the momentum of the scattered electron of $ns[3/2]_1$, while $f_{ns'[1/2]_1}$, E_2 and p_2 are the corresponding quantites of $ns[3/2]_1$, respectively.

From Eq. (10) it can be seen that the GOS ratio for the pair of transitions, which are formed by the same electronic configuration through the intrachannel interaction, should be a constant. Now the parallel character of the GOSs of $(n-1)p^5ns[3/2]_1$ and $(n-1)p^5ns'[1/2]_1$ of Ne (n=3), Ar (n=4), and Kr (n=5) shown in Figs. 3, 5 and 6 can be easily elucidated: the *LS* coupling singlet nature of the ground states of noble atoms selects the *LS* coupling singlet components from the intermediate coupling wave functions of the excited states, and the *LS* coupling singlet components are enough to describe the excitation process from the ground state for fast electron impact. The quantitative GOS ratios for $(n-1)p^5ns[3/2]_1$ and $(n-1)p^5ns'[1/2]_1$ of Ne (n=3), Ar (n=4), and Kr (n=5) are shown in Fig. 7, and it can be seen that the agreement between the experimental results and the theoretical ones are satisfactory.

Recently, the GOS ratios for the higher dipole-allowed excitations of Ne [19], the dipole-forbidden excitations of kr [34], as well as the inner shell excitations of Li [38] shown in Fig.8, were measured. All these results [19, 34, 38] show that the GOSs ratios for the pair of transitions formed by the same electronic configuration through the intrachannel interaction are constant, which is consistent with the above theoretical analysis. Furthermore, since the photoexcitation and photoionization are the electric dipole processes, which correspond to the excitation and ionization for fast electron impact at optical limit ($K^2 \to 0$), the 4-parameter model, which depends on the validity of *LS* coupling, suggested by Snell *et al* [39] is sufficient to describe the Xe 4d photoionization process.

5. Conclusion

The DCSs and apparent GOSs for the valence-shell excitations of the noble atoms of He, Ne, Ar, and Kr measured by the intermediate and high energy impact are reviewed, and the conditions that the FBA holds for the excitations are discussed. The asymptotical behaviors of the apparent GOSs of He and Ne show that the FBA is more easily reached for the dipole-allowed transition than the dipole-forbidden one of $^1S_0 \to ^1 S_0$ type, in which the final state has the same term symbol of the initial state. From the measured GOS curves of $(n-1)p^5ns[3/2]_1$ and $(n-1)p^5ns'[1/2]_1$ of Ne (n=3), Ar (n=4), and Kr (n=5), the parallel character of the pair transitions was observed. With the help of the calculated intermediate coupling coefficients, this phenomenon is elucidated by the fact that the *LS* coupling singlet nature of the ground state selects the *LS* coupling singlet component from the excited wave function, i.e., the contribution from the singlet component in the excitation is dominant in the generalized oscillator strength. Furthermore, the similar dynamic behavior was observed not only for the valence-shell excitations but also for the inner-shell excitations and the dipole-forbidden excitations.

Acknowledgments

The author acknowledges the supports by the National Nature Science Foundation of China (Grants Nos. 10734040 and 10874168), CAS Knowledge Promotion Project (Grants No. KJCS1-YW-N30) and National Basic Research Program of China (Grant No. 2010CB923301).

References

[1] Bethe H 1930 *Ann. Phys.* **5** 325
[2] Inokuti M 1971 *Rev. Mod. Phys.* **43** 297
[3] Cartwright D C, Csanak G, Trajmar S and Register D F 1992 *Phys. Rev. A* **45** 1602
[4] Trajmar S, Register D F, Cartwright D C and Csanak G 1992 *J. Phys. B: At. Mol. Opt. Phys.* **25** 4889
[5] Dillon M A and Lassettre E N 1975 *J. Chem. Phys* **62** 2373
[6] Dillon M A 1975 *J. Chem. Phys* **63** 2035
[7] Suzuki T Y, Suzuki H, Currell F J, Ohtani S, Sakai Y, Takayanagi T and Wakiya K 1998 *Phys. Rev. A* **57** 1832
[8] Sakai Y, Suzuki T Y, Min B S, T Takayanagi K W, Suzuki H, Ohtani S and Takuma H 1991 *Phys. Rev. A* **43** 1656
[9] Suzuki T Y, Suzuki H, Ohtani S, Takayanagi T and Okada K 2007 *Phys. Rev. A* **75** 032705
[10] Xu K Z, Feng R F, Wu S L, Ji Q, Zhang X J, Zhong Z P and Zheng Y 1996 *Phys. Rev. A.* **53** 3081

[11] Zhong Z P, Feng R F, Wu S L, Zhu L F, Zhang X J and Xu K Z 1997 *J. Phys. B: At. Mol. Opt. Phys.* **30** 5305
[12] Liu X J, Zhu L F, Yuan Z S, Li W B, Cheng H D, Sun J M and Xu K Z 2004 *J. Electron Spectrosc. Relat. Phenom.* **135** 15
[13] Zhu L F, Zhong Z P, Liu X J, Feng R F, Zhang X J and Xu K Z 1999 *J. Phys. B: At. Mol. Opt. Phys.* **32** 4897
[14] Han X Y and Li J M 2006 *Phys. Rev. A.* **74** 062711
[15] Cann N M and Thakkar A J 2002 *J. Electron Spectrosc. Relat. Phenom.* **123** 143
[16] Fursa D V and Bray I 1995 *Phys. Rev. A* **52** 1279
[17] Xie B P, Zhu L F, Yang K, Zhou B, Hiraoka N, Cai Y Q, Yao Y, Wu C Q, Wang E L and D L Feng 2009 *arXiv:0906.3210*
[18] Brunger M J and Buckman S J 2002 *Phys. Rep.* **357** 215
[19] Cheng H D, Zhu L F, Yuan Z S, Liu X J, Sun J M, Jiang W C and Xu K Z 2005 *Phys. Rev. A.* **72** 012715
[20] Khakoo M A, Wrkich J, Larsen M, Kleiban G, Kanik I, Trajmar S, Brunger M J, Teubner P J O, Crowe A, Fontes C J, Clark R E H, Zeman V, Bartschat K, Madison D H, Srivastava R and Stauffer A D 2002 *Phys. Rev. A* **65** 062711
[21] Khakoo M A, Vandeventer P, Childers J G, Kanik I, Fontes C J, Bartschat K, Zeman V, Madison D H, Saxena S, Srivastava R and Stauffer A D 2004 *J. Phys. B: At. Mol. Opt. Phys.* **37** 247
[22] Suzuki T Y, Suzuki H, Ohtani S, Min B S, Takayanagi T and Wakiya K 1994 *Phys. Rev. A* **49** 4578
[23] Li G P, Takayanagi T, Wakiya K, Suzuki H, Ajiro T, Yagi S, Kano S S and Takuma H 1988 *Phys. Rev. A* **38** 1240
[24] Takayanagi T, Li G P, Suzuki H, Ajiro T, Kano S S and Takuma H 1990 *Phys. Rev. A* **41** 5948
[25] Ji Q, Wu S L, Feng R F, Zhang X J, Zhu L F, Zhong Z P, Xu K Z and Zheng Y 1996 *Phys. Rev. A* **54** 2786
[26] Zhu L F, Cheng H D, Yuan Z S, Liu X J, Sun J M and Xu K Z 2006 *Phys. Rev. A.* **73** 042703
[27] Zhu L F, Yuan H, W C Jiang F X Z, Yuan Z S, Cheng H D and Xu K Z 2007 *Phys. Rev. A.* **75** 032701
[28] Fan X W and Leung K T 2000 *Phys. Rev. A* **62** 062703
[29] Li W B, Zhu L F, Liu X J, Yuan Z S, Sun J M, Cheng H D, Zhong Z P and Xu K Z 2003 *Phys. Rev. A* **67** 062708
[30] Zhu L F, Zhang F X, Cheng H D, Yuan H, Yuan Z S, Li W B, Liu X J and Xu K Z 2007 *Chin. Phys. Lett.* **16** 2938
[31] Amusia M Y, Chernysheva L V, Felfli Z and Msezane A Z 2003 *Phys. Rev. A* **67** 022703
[32] Amusia M Y, Chernysheva L V, Felfli Z and Msezane A Z 2007 *Phys. Rev. A* **75** 062703
[33] Chen Z F and Msezane A Z 2000 *J. Phys. B: At. Mol. Opt. Phys.* **33** 5397
[34] Cheng H D, Zhu L F, Liu X J, Yuan Z S, Li W B and Xu K Z 2005 *Phys. Rev. A.* **71** 032714
[35] Platzman P M and Tzoar N 1965 *Phys. Rev.* **139** A410
[36] Eisenberger P and Platzman P M 1970 *Phys. Rev. A* **2** 415
[37] Amusia M Y, Chernysheva L V, Felfli Z and Msezane A Z 2002 *Phys. Rev. A* **65** 062705
[38] Jiang W C, Zhu L F and Xu K Z 2008 *Chin. Phys. Lett.* **25** 3649
[39] Snell G, Drescher M, Muller N, Heinzmann U, Hergenhahn U, Viefhaus J, Heiser F, Becker U and Brookes N B 1996 *Phys. Rev. Lett.* **76** 3923

Silicon thermal oxidation and its thermal desorption investigated by Si 2p core-level photoemission

Y. Enta, H. Nakazawa, S. Sato, H. Kato, and Y. Sakisaka

Graduate School of Science and Technology, Hirosaki University
3 Bunkyo-cho, Hirosaki 036-8561, Japan

E-mail: enta@cc.hirosaki-u.ac.jp

Abstract. Initial thermal oxidation of silicon and thermal desorption of silicon oxide have been investigated by Si 2p core-level photoemission with synchrotron radiation. The surface reaction processes are discussed from the difference in the chemically-shifted components, especially suboxides. On the thermal oxidation, time evolutions of suboxide intensities show a distinct temperature dependence, which is explained by two growth modes for the oxidation: first-order Langmuir-type adsorption mode and two-dimensional island growth mode at oxidation temperatures below and above 650 °C, respectively. On the thermal desorption, the spectrum for the thermal oxide followed by annealing at 1000 °C for 30 s in vacuum, whose thickness is nominally estimated at 3.7 Å, are compared with the 3.3 Å-thickness thermal oxide without the process of the annealing. The intensities of the suboxide components are considerably different between them, which can be explained by the formation and lateral growth of the voids in the annealed oxide layer.

1. Introduction

Since the silicon gate oxide in highly integrated metal-oxide-semiconductor field-effect-transistor (MOS-FET) devices is now required to reduce the thickness to a few tens of angstroms, understanding on the oxidation kinetics at the initial stage has become one of the most important issues in the semiconductor industry. In this respect, photoelectron spectroscopy for Si 2p core level provides useful information on the surface reaction, because with synchrotron radiation at optimized energy, it has high surface sensitivity of a few angstroms. Since the silicon-dioxide (SiO_2) component has large chemical shift about 4 eV in the Si 2p core-level spectrum and thus can be apparently distinguished from silicon-substrate component, it is easy to measure the oxidation rate at the initial stage. In addition, analysis of intermediated oxidation states, so-called suboxides [1,2], gives valuable information on atomic structures at SiO_2/Si interface.

In the previous studies, we measured the time evolutions of the SiO_2 component in the Si 2p core-level spectrum during the initial thermal oxidation of Si(100). We concluded that, at temperatures below 650 °C, the oxidation proceeds randomly without formation of two-dimensional (2D) islands by the observation that the growth rate follows a first-order Langmuirian behavior, while at temperatures above 650 °C, the oxidation was indicated to proceed via 2D nucleation because the growth exhibits some incubation time [3,4]. As mentioned above, the suboxide components in the spectrum, which exist at SiO_2/Si interface, are also very informative. In this study, therefore, we focus on the temperature dependence of the suboxide components to investigate the surface reaction at the initial stage and verify the above oxidation behaviors.

The thermal-desorption process of the SiO_2 layer on Si is also crucial in MOS-FET technology. Tromp et al. reported that 100 Å-thickness SiO_2 layer is thermally removed from the Si surface by the formation and lateral growth of voids in the oxide, exposing regions of atomically clean Si, while the surrounding oxide retains its initial thickness [5]. The same desorption behaviors were also observed for thinner oxide layer [6-8]. We here demonstrate the formation of the same kind of void by a comparison of the suboxide Si-2p components for thermal oxide layers with and without the process of the thermal desorption.

2. Experimental

Experiments were carried out at the beam line BL-3B at the Photon Factory, the High Energy Accelerator Research Organization. The photon energy was set at 135 eV, where the photoelectron inelastic attenuation length from the Si 2p core level in silicon is minimized. A mirror-polished, B-doped Si(100) wafer cut to a size of 20×5×0.5 mm^3 was used as a silicon substrate. The wafer was chemically treated to passivate the surface and then annealed by resistive heating at 1000 °C in ultrahigh vacuum. The cleanliness of the surface was checked by the surface state in valence-band and Si 2p core-level spectra, which showed no contamination. For thermal oxidation, the 99.99 % oxygen gas was introduced into a reaction chamber through a variable leak valve. The overall instrumental energy resolution was estimated at about 0.2 eV. The Si 2p core-level spectra are deconvoluted by a least-square fitting procedure using the spin-orbit split Voigt functions.

3. Results and Discussion

3.1. Thermal oxidation

Figure 1 shows typical Si 2p core-level spectra of oxides on Si(100) grown at an oxygen pressure of 10^{-6} Torr and oxidation temperatures of (a) 700 °C and (b) 500 °C. All spectra consist of seven components: substrate silicon (Si^0), three suboxides (Si^{1+}, Si^{2+}, Si^{3+}), stoichiometric oxide, that is, SiO_2 (Si^{4+}), and two associated with surface strain layer [9] and/or bare Si surface [10] (S1 and S2). The hatched peaks show the components at the oxidation time of 1 min, whose intensities are normalized by the intensity of Si^0 in figure 2(a). We can see a distinct difference in the intensities of the suboxide components, Si^{2+} and Si^{3+}, as well as S1 component between the spectra at 700 °C and 500 °C, although the intensity of Si^{4+} is almost same. Namely, the intensities of Si^{2+} and Si^{3+} at 700 °C are much smaller than those at 500 °C, while the intensity of S1 shows opposite behavior. Figures 2 (b) and 2 (c) show oxidation-time dependences of the suboxide intensities for 700 °C and 500 °C, respectively, normalized by Si^{4+}. At the temperature of 700 °C, both Si^{2+} and Si^{3+} intensities are almost kept constant, which means they increase in proportion to Si^{4+}. On the other hand, at the temperature of 500 °C, as the oxidation proceeds, they both decrease monotonically.

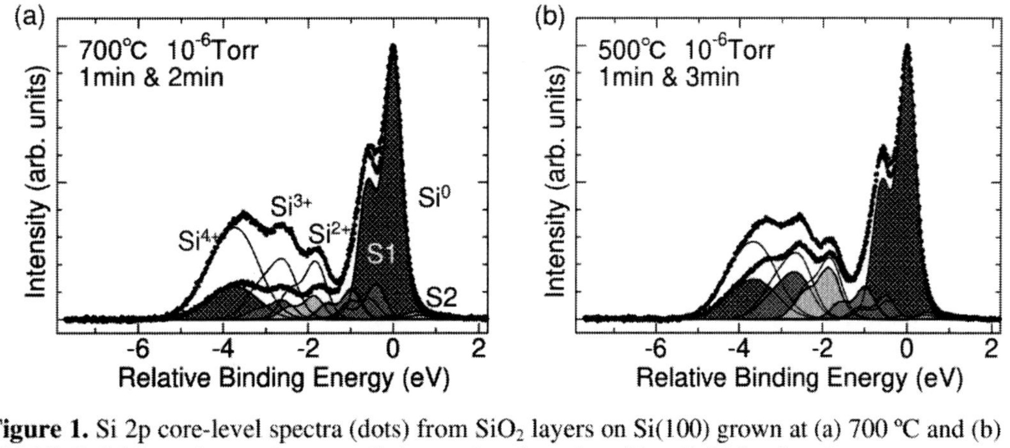

Figure 1. Si 2p core-level spectra (dots) from SiO_2 layers on Si(100) grown at (a) 700 °C and (b) 500 °C. Solid lines denote the curve fittings using the spin-orbit split Voigt functions.

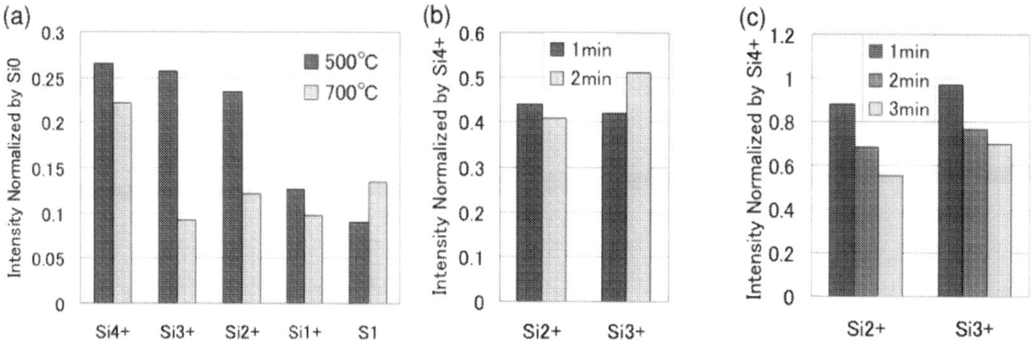

Figure 2. Comparisons of intensity ratios of chemical-shift components for Si 2p core-level spectra from SiO_2 layer grown (a) for 1 min, (b) at 700 °C, and (c) at 500 °C.

We interpret these results as the temperature dependence of the oxidation kinetics. As mentioned in the introduction, in the high temperature regime, the oxidation proceeds via 2D nucleation, which is probably due to SiO_2 formation, because the oxygen atom adsorbed on the surface is not stable and thus quickly desorbs from the surface as a volatile SiO molecule [11,12]. Once SiO_2 nucleus is formed, it laterally expands on the surface. Therefore, the suboxides exist only at the SiO_2/Si interface as shown in figure 3 (a). On the other hand, in the low temperature regime, the suboxides gradually change to SiO_2 as shown in figure 3 (b) because no SiO desorption occurs at the low temperature, showing Langmuirian behavior. Since the suboxides exist not only at the SiO_2/Si interface but also on the substrate Si surface, it should be observed that the suboxide intensities at 500 °C are higher than those at 700 °C as shown in Fig. 2 (a). With proceeding of the oxidation, it is obvious that the intensity ratio of suboxide to Si^{4+} decrease monotonically in case of figure 3 (b). In case of figure 3 (a), on the other hand, it is kept constant because SiO_2 island laterally grows without change of its thickness. This inference is consistent with the results in figure 2 (b) and (c). Thus, the oxidation kinetics mentioned above is reconfirmed by the observation of the change in the suboxide intensity in the Si 2p core-level spectra.

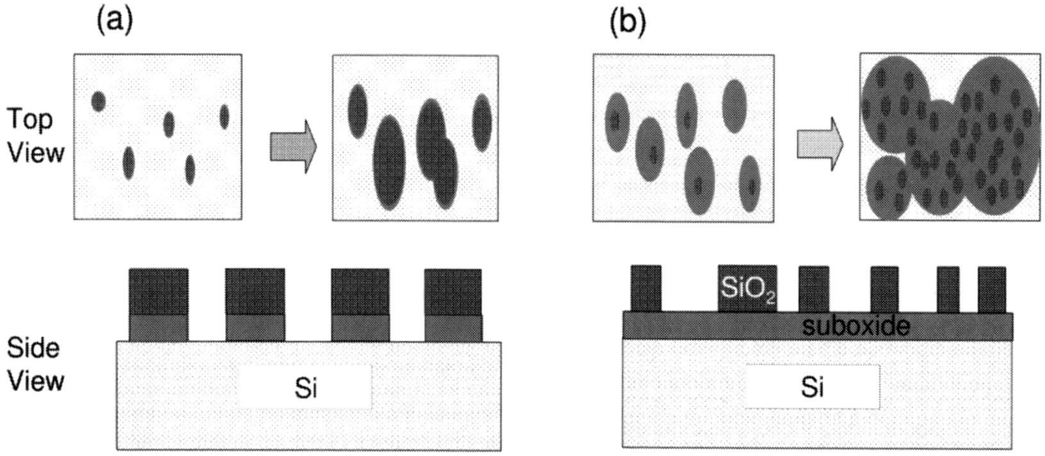

Figure 3. Schematic models for oxidation processes grown at (a) 700 °C and (b) 500 °C.

3.2. Thermal desorption

To investigate the thermal desorption of the SiO_2 layer, we first formed a thicker oxide layer. Figure 4(a) shows Si 2p core-level spectrum of the surface oxidized for 10 min at an oxygen pressure of 1 Torr and oxidation temperature of 800 °C. The thickness of the SiO_2 layer (D) can be calculated from

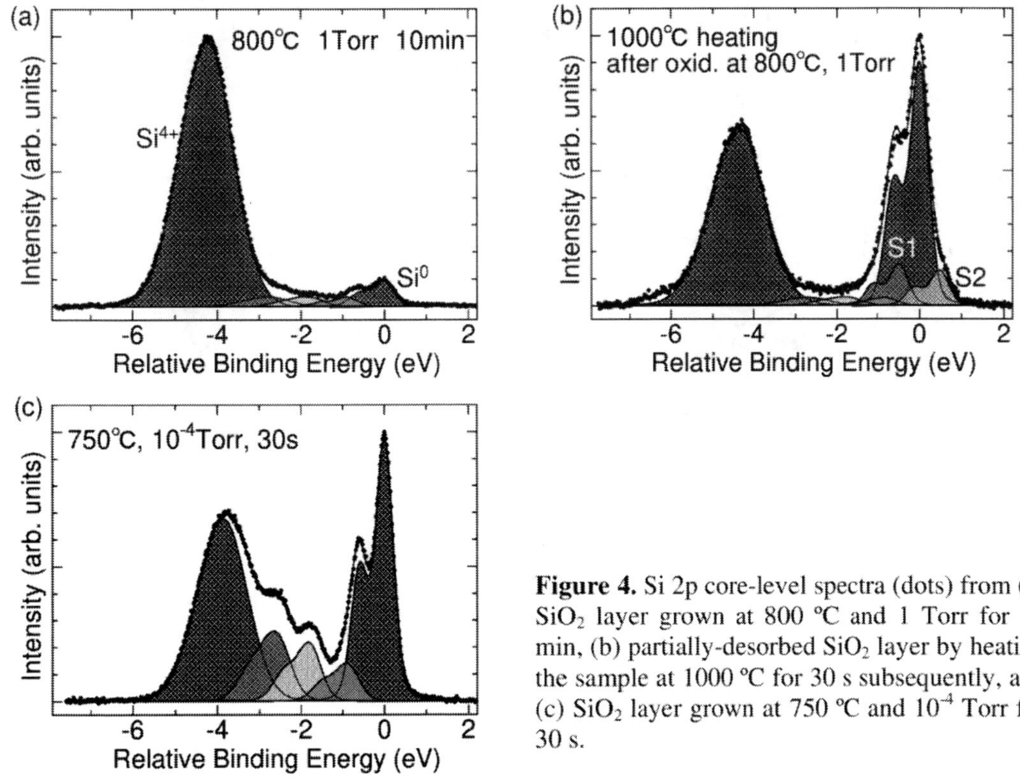

Figure 4. Si 2p core-level spectra (dots) from (a) SiO_2 layer grown at 800 °C and 1 Torr for 10 min, (b) partially-desorbed SiO_2 layer by heating the sample at 1000 °C for 30 s subsequently, and (c) SiO_2 layer grown at 750 °C and 10^{-4} Torr for 30 s.

the formula [13,14]:

$$D = \lambda_{SiO_2} \sin\theta_e \ln\left(\frac{I_{Si^{4+}} I_0}{I_{Si^0} I_\infty} + 1 \right) \qquad (1)$$

$$I_0 = n_{Si}\sigma_{Si}\lambda_{Si}$$

$$I_\infty = n_{SiO_2}\sigma_{SiO_2}\lambda_{SiO_2}$$

where $I_{Si^{4+}}$ is the measured Si^{4+} intensity, I_{Si^0} is the measured Si^0 intensity, θ_e is the electron emission angle with respect to the surface, λ_Z is the electron inelastic attenuation length in species Z (Z=Si or SiO_2), n_Z is density of Si atoms in species Z, and σ_Z is the relevant differential atomic photo-ionization cross section of the Si 2p core level in Z. At a photon energy of 135 eV, λ_{SiO_2} and the ratio I_0/I_∞ are estimated from previous analyses of experimental data [13,14] at 0.68 nm and 0.5, respectively. Using these values, the SiO_2 thickness in figure 4 (a) is estimated at 15.0 Å. After the oxidation, the sample was heated at the temperature of 1000 °C for 30 s to desorb a part of the SiO_2 layer, which is shown in Fig. 4(b). The SiO_2 thickness for this sample is estimated at 3.7 Å, if we simply use the intensity ratio of Si^{4+} to Si^0 for equation (1). However, the spectrum in figure 4 (b) is quite different from that for the thermal oxide layer without the process of the thermal desorption shown in figure 4 (c). The SiO_2 layer in figure 4 (c) is thermally grown at the temperature of 750 °C and at the pressure of 10^{-4} Torr for 30 s, whose thickness is estimated at 3.3 Å. In spite of the same intensity ratio of Si^{4+} to Si^0, it is obviously seen that the suboxide intensities in figure 4 (b) are much less than those in figure 4 (c). The SiO_2 layer in figure 4 (c) is expected to be homogeneously grown as shown in figure 5 (c). In order to explain the suboxide intensities in figure 4 (b), therefore, we have to consider the SiO_2 layer spatially inhomogeneous as shown in figure 5 (b). Namely, the SiO_2 layer is desorbed from the Si surface by the formation of voids in the SiO_2 layer, where a bare Si surface is exposed. In such a surface morphology, the intensity ratio of Si^{4+} to Si^0 becomes small owing to the contribution to the Si^0 component from the bare Si surface even if the SiO_2 thickness is unchanged. The formation of the voids is also confirmed by the appearance of S1 and S2 components due to the bare Si surface in figure 4 (b), where the contribution by the surface strain layer in the oxide regime can be neglected owing to the thick oxide overlayer.

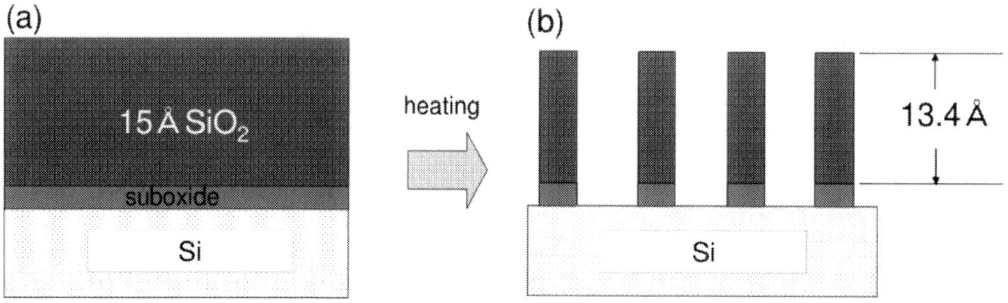

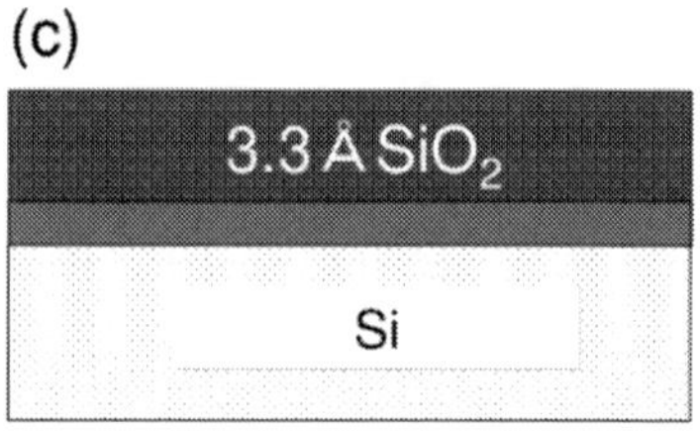

Figure 5. Schematic of surface structures for SiO_2 layers shown in figure 4.

If the SiO_2 thickness is kept constant during the desorption process ((a) to (b)), the intensity ratios of suboxide to Si^{4+} should be unchanged. The intensities of Si^{X+}/Si^{4+} (X=2 and 3) in figure 4 (a) are 2.60×10^{-2} and 2.76×10^{-2}, respectively, and those in figure 4 (b) are 3.51×10^{-2} and 3.60×10^{-2}, respectively. The small increase of these values after the sample heating indicates the small decrease of the SiO_2 thickness. When the suboxide exists only at the SiO_2/Si interface as shown in figure 5, the intensity ratio of Si^{X+} to Si^{4+} is given by

$$\frac{I_{Si^{X+}}}{I_{Si^{4+}}} = \frac{n_Z \sigma_Z \lambda_Z \{1-\exp(-A/\lambda_Z)\}}{n_{SiO_2}\sigma_{SiO_2}\lambda_{SiO_2}\{\exp(D/\lambda_{SiO_2})-1\}} \tag{2}$$

where A is the suboxide thickness and Z stands for the Si^{X+} suboxide. From this equation, the SiO_2 thickness in figure 4 (b) is estimated at 13.4 Å, which is only a 10 percent reduction, if the thickness of the suboxide layer at the interface is unchanged by the sample heating. In the case in figure 4 (b), the intensity ratio of Si^{4+} to Si^0 is given by

$$\frac{I_{Si^{4+}}}{I_{Si^0}} = \frac{I_\infty}{I_0}\frac{C\{1-\exp(-D/\lambda_{SiO_2})\}}{(1-C)\exp(-S/\lambda_{Si})+C\exp(-D/\lambda_{SiO_2})} \tag{3}$$

where C is the surface coverage of the SiO_2 layer, and S is the thickness of the surface layer of bare Si corresponding to the S1 and S2 components, which involve the second layer and the surface defect [10] in our spectra. Therefore, we obtain

$$C = \frac{1}{1-\exp(S/\lambda_{Si}-D/\lambda_{SiO_2})+I_{Si^0}/I_{Si^{4+}}\cdot I_\infty/I_0\{\exp(S/\lambda_{Si})-\exp(S/\lambda_{Si}-D/\lambda_{SiO_2})\}} \ . \tag{4}$$

Using the value of 13.4 Å calculated by equation (2) as *D*, *C* is estimated at 0.38, when S is assumed to be 3 Å, which corresponds to a few atomic layers of Si(100). Although further studies are indispensable to justify these estimations, the analysis of Si 2p core-level spectra gives valuable information on the desorption process of the SiO_2 layer.

4. Conclusion

Initial thermal oxidation of silicon and thermal desorption of silicon oxide have been investigated by Si 2p core-level spectra, especially suboxide components. The behaviors of the suboxide intensities clearly show temperature dependence, which is explained by two growth modes for the oxidation: a Langmuir-type adsorption mode and 2D island growth mode at temperatures of 500 °C and 700 °C, respectively. The SiO_2 layer partially desorbed by heating the sample at 1000°C indicates the inhomogeneous surface morphology, which is consistent with the void-formation model. In addition, the oxide thickness and the oxide coverage are estimated by the chemical-shift-component intensities of the SiO_2 layer.

Acknowledgement

This work was partly supported by the Grant-in-Aid for Scientific Research of the Ministry of Education, Culture, Sports, Science and Technology of Japan (Nos. 215603219 and 20656007), the Grant for Priority Research Designated by the President of Hirosaki University, and Iketani Science and Technology Foundation (No. 0201007). This work was performed under the approval of the Photon Factory Program Advisory Committee (Proposal Nos. 2006G353 and 2008G673).

References

[1] Hattori T and Suzuki T 1983 *Appl. Phys. Lett.* **43** 470
[2] Hollinger G and Himpsel F J 1983 *Phys. Rev.* B **28** 3651
[3] Enta Y, Takegawa Y, Suemitsu M and Miyamoto N 1996 *Appl. Surf. Sci.* **100/101** 449
[4] Enta Y, Miyanishi Y, Irimachi H, Niwano M, Suemitsu M, Miyamoto N, Shigemasa E and Kato H 1998 *Phys. Rev.* B **57** 6294
[5] Tromp R, Rubloff G W, Balk P, LeGoues F K and Loenen E J 1985 *Phys. Rev. Lett.* **55** 2332
[6] Sun Y -K, Bonser D J and Engel T 1991 *Phys. Rev.* B **43** 14309
[7] Johnson K E and Engel T 1992 *Phys. Rev. Lett.* **69** 339
[8] Takakuwa Y, Nihei M, Horie T and Miyamoto N 1994 *J. Non-Cryst. Solids* **179** 345
[9] Oh J H, Yeom H W, Hagimoto Y, Ono K, Oshima M, Hirashita N, Nywa M, Toriumi A and Kakizaki A 2001 *Phys. Rev.* B **63** 205310
[10] Landemark E, Karlsson C J, Chao Y -C and Uhrberg R I G 1992 *Phys. Rev. Lett.* **69** 1588
[11] Tabe M, Chiang T T, Lindau I and Spicer W E 1986 *Phys. Rev.* B **34** 2706
[12] Luts F, Bischoff J L, Kubler L and Bolmont D 1989 *Phys. Rev.* B **40** 10356
[13] Himpsel F J, McFreely F R, Taleb-Ibrahimi A, Yarmoff J A and Hollinger G 1988 *Phys. Rev.* B **38** 6084
[14] Seah M P and Spencer S J 2002 *Surf. Interface Anal.* **33** 640

Imaging magnetic atoms by Photoelectron Diffraction

Takahisa Ishiyama*, Masahiko Wada, Kouta Ito, Takehisa Konishi, Takashi Fujikawa†

*,†Graduate School of Advanced Integration Science, Chiba University, 1-33 Yayoi-cho, Inage-ku, Chiba 263-8522, Japan

E-mail: spin@graduate.chiba-u.jp

E-mail: tfujikawa@faculty.chiba-u.jp

Abstract. We study possible methods how to obtain imaging of magnetic atoms by use of photoelectron diffraction (PD). We have proposed a method to apply Daimon effect where forward focusing PD peaks are rotated; we use the difference of the PD intensities for different circular polarization but $-\phi$ is used for the $-$ circular polarization only in the data handling. This technique allows us to obtain atomic image only of spin polarized atoms. In this work we further apply this technique to ferromagnetic and antiferromagnetic surface layers with in-plane magnetization adsorbed on a nonmagnetic layer.

1. Introduction

Bansmann *et al.* have observed circular dichroism in angular distribution (CDAD) of photoelectrons excited from spherically symmetric initial states of nonmagnetic species such as CO/Pd(111) [1]. Daimon *et al.* have observed strong CDAD excited from Si 2p core level on Si(001) surface, although this system is not chiral [2]. They proposed an interesting formula for the azimuthal shift $\Delta\phi$ from the forward focusing peak. In order to study these effects, some theoretical approaches have successfully been proposed [3–6] based on short-range-order multiple scattering theory. By using CDAD effects, Daimon has proposed a new technique to observe stereo photographs of surface atomic arrangements [7–9].

In order to extract only information on magnetic atoms, we apply the cancellation of rotated forward focusing peaks because of the Daimon effect for nonmagnetic atoms. Conventional CDAD is defined by the difference $\Delta\tilde{I}(\theta, \phi) = I^{+}(\theta, \phi) - I^{-}(\theta, \phi)$ for different circular polarization $\pm$. This CDAD is influenced by surrounding magnetic and nonmagnetic atoms. We thus rather use the modified CDAD $\Delta I(\theta, \phi) = I^{+}(\theta, \phi) - I^{-}(\theta, -\phi)$. This technique allows us to obtain atomic image only of spin polarized atoms, and to distinguish magnetic atoms with up spins from those with down spins; magnetization is assumed to be parallel to the X-ray incidence [10]. For this purpose the scattering atoms should be arranged with mirror symmetry in regard to the xz-plane. The atom at ϕ_α is transformed into the atom at $-\phi_\alpha$ under the mirror symmetry operation. We thus expect the cancellation in the modified CDAD $\Delta I(\theta, \phi)$ when they are nonmagnetic equivalent atoms. In this we have specified the x-axis.

In this paper we furthermore apply this technique to ferromagnetic and antiferromagnetic surface layers with in-plane magnetization adsorbed on a nonmagnetic layer. The result demonstrates potential use of this technique.

2. Theory

X-ray photoelectron diffraction (XPD) intensity $I_{\pm}^{j_c}(\mathbf{k},\sigma)$ excited from the splitted subshell j_c (=3/2) measuring photoelectrons with momentum $\boldsymbol{k}$ and spin σ is written for the $\pm$ X-ray circular polarization [11]

$$I_{\pm}^{j_c}(\mathbf{k},\sigma) = |\sum_{\alpha}\sum_{\mu_c=-j_c}^{j_c} e^{-\kappa d_\alpha(\hat{\mathbf{k}})}e^{-i\mathbf{k}_r\cdot\mathbf{R}_{\alpha A}}\sum_{LL'} Y_{L'}(\hat{\mathbf{k}})[(1-X)^{-1}]_{L'L}^{\alpha A} M_{LL_c}^{\pm}(\mathbf{k};\sigma)|^2 \quad (1)$$

$$X_{LL'}^{\alpha\beta} = t_l^{\alpha\sigma}(k_r)G_{LL'}(k_r R_{\alpha\beta})e^{-\kappa R_{\alpha\beta}}(1-\delta^{\alpha\beta}) \quad (2)$$

where X is a spin-dependent square matrix, where matrix elements are labeled by atomic sites (A, α, β, ...) and angular momenta L. $d_\alpha(\hat{\mathbf{k}})$ is the distance from site α to the surface of the solid along the direction of a photoelectron propagation $\hat{\mathbf{k}}$, $\mathbf{R}_{\alpha A}$ is the position vector of the scatterer α measured from photoelectron emitter A.

In the dipole approximation, the photoexcitation matrix element $M_{LL_c}^{\pm}(\mathbf{k};\sigma)$ excited by circularly polarized light ($\mathbf{k}$ // z) is given by [12, 13]

$$M_{LL_c}^{\pm}(\mathbf{k};\sigma) = \sqrt{\frac{2}{\pi}} i^{-l} e^{i\delta_l^{A\sigma}} \rho(l)_c G(L_c 1 \pm 1|L)\left\langle L_c \frac{1}{2}\frac{\sigma}{2}|j_c\mu_c\right\rangle, \quad (3)$$

$$\rho(l)_c = \int R_l(kr)R_{l_c}(r)r^3 dr, \quad (4)$$

where $\delta_l^{A\sigma}$ is the spin-dependent phase shift of l th partial wave at site A, and $R_l(kr)$ and $R_{l_c}(r)$ describe the radial parts of the l th partial wave and the core function. Gaunt integral $G(L_c 1 \pm 1|L) = \int Y_{l_c}(\mathbf{r})Y_{1\pm1}(\mathbf{r})Y_L^*(\mathbf{r})d\hat{\mathbf{r}}$ is responsible for the angular momentum selection rule of the photoexcitation. The angular momentum representation of the site-t matrix $t_l^{\alpha\sigma}(k)$ at site α is given by

$$t_l^{\alpha\sigma}(k) = -\frac{e^{2i\delta_l^{\alpha\sigma}}-1}{2ik} \quad (5)$$

in terms of the phase shift $\delta_l^{\alpha\sigma}$ at site α for spin σ and the photoelectron wave number k.

We introduce modified CDAD which enables us to pick up only the information on magnetic atoms [10],

$$\Delta I(\theta,\phi) = I^+(\theta,\phi) - I^-(\theta,-\phi). \quad (6)$$

3. Results

Here we show some calculated results for perpendicular and in-plane magnetization models.

3.1. Cu-Gd_4 model with perpendicular magnetization

First we consider perpendicularly magnetized Cu-Gd_4 model, where circular polarized X-ray photons propagates parallel to z-axis. We consider photoemission from Cu $2p_{3/2}$ sublevel with kinetic energy of 100 eV (ϵ_k = 100 eV). We assume that Cu atoms are nonmagnetic whereas a Gd atom is strongly magnetized S= 7/2 which aligns parallel or antiparallel to the z direction. We consider the following three models; (a) ferro model, (b) antiferro modell in which two Gd atoms have down spin at α and γ sites, and (c) antiferro modle2 in which two Gd atoms have down spin at α and β sites.

Figure 1 shows the calculated modified circular dichroism ΔI(θ,ϕ) for the different spin orders. Figure 1(a) shows four positive spots corresponding to forward focusing Cu→Gd (up spin) scatterings. Figure 1(b) shows two positive spots corresponding to the forward Cu→Gd (up spin at β and δ) scatterings, and two negative spots to Cu→Gd (down spin at α and γ) scatterings. This approach can clearly distinguish ferromagnetic order from antiferromagnetic order [10]. We also consider the other antiferromagnetic model (c) where α and β Gd atoms are magnetized with down spin and γ and δ with up spin. The calculated modified circular dichroism is shown by (c) where α and γ are in xz-plane, and is shown by (c)' where x axis is positively rotated around z-axis. In (c) the mirror symmetry in regard to xz-plane is lost. Only the spots for α and γ which are in the xz-plane are found, and spots for β and δ disappear. In (c)' the mirror symmetry in regard to the new xz-plane correctly gives rise to magnetic atom imaging. When we rotate the xy-plane with $-45°$, we obtain perfectly green pattern for $\Delta I(\theta, \phi)$: We find no spot. We thus understand the importance of the mirror symmetry in regard to xz-plane.

In this paper we use the full multiple scattering calculations as given by eq.(1). In the previous paper the plane wave approximation was used at the same energy, $\epsilon_k = 100$ eV. The result is quite similar to each other, where four positive spots are found for the ferromagnetic order in the plane wave approximation [10].

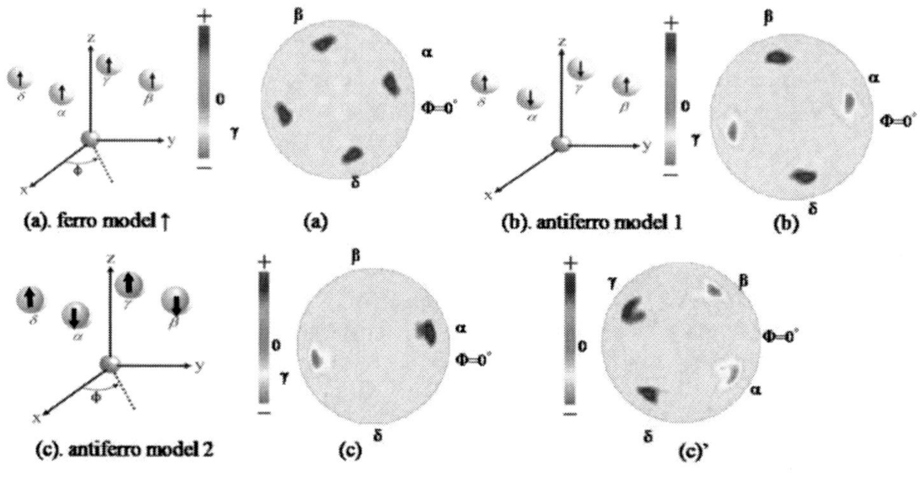

Figure 1. Left: Cu-Gd$_4$ models magnetized perpendicular to the surface: (a) ferro, (b) antiferro1 (α and γ are magnetized with down spin), (c) antiferro2 (α and β are magnetized with down spin). Right: Calculated modified CDAD $\Delta I(\theta, \phi)$. The calculated $\Delta I(\theta, \phi)$ are shown for the model (c) by (c) and (c)', where α and γ are in the xz-plane, and xy-plane is rotated with 45° around z-axis. In (c) we have no mirror symmetry with respect to xz-plane, whereas in (c)' we have the symmetry.

3.2. Cu-Gd$_4$ model with in plane magnetization

Next we consider in-plane magnetized Cu-Gd$_4$ model where Gd has S= 7/2 aligned parallel or antiparallel to the x direction with same photoelectron kinetic energy of 100 eV ($\epsilon_k = 100$ eV). Figure 2 shows $\Delta I(\theta, \phi)$ for the different spin orders (a) ferro model in which all Gd atoms are ordered parallel to x direction, (b) antiferro model1 which has two Gd with spin parallel to x direction at α and γ sites, and (c) antiferro model2 which has two Gd with spin parallel to x direction at α and β sites: Other two Gd atoms are magnetized in opposite direction.

Figure 2(a) shows the two negative and positive spots corresponding to Cu→Gd forward scatterings even for the ferromagnetic model. The inversion symmetry in regard to the center observed in Figs. 2(a) and (b) is explained by the symmetry consideration. The mirror symmetry with respect to xz-plane gives rise to spin inversion of the spin x component S_x, $S_x \to -S_x$, and $Y_{11} \to Y_{1-1}$. The focusing spot near α site is approximately written

$$\Delta I(\alpha) = I^{+}_{x-}(\alpha) - I^{-}_{x-}(\alpha).$$

Rotation around z-axis with 180° gives by noting that $S_x \to -S_x$, and $Y_{1\pm1} \to Y_{1\pm1}$

$$\begin{aligned} \Delta I(\gamma) &= I^{+}_{x-}(\gamma) - I^{-}_{x-}(\gamma) \\ &= I^{+}_{x+}(\alpha) - I^{-}_{x+}(\alpha). \end{aligned} \tag{7}$$

Mirror symmetry operation in regard to xz-plane yields a relation

$$\Delta I(\gamma) = I^{-}_{x-}(\alpha) - I^{+}_{x-}(\alpha). \tag{8}$$

Comparing eqs.(7) and (8), we obtain a relation between $\Delta I(\alpha)$ and $\Delta I(\gamma)$

$$\Delta I(\alpha) = -\Delta I(\gamma). \tag{9}$$

In the same way, we can show that $\Delta I(\beta) = -\Delta I(\delta)$ in the model (a). That is also the case for the antiferromagnetic model (b). When the lack of the mirror symmetry as in (c), two spots corresponding to β and δ are missing in the present imaging technique. In the imaging (c)' we find four spots because of the mirror symmetry. In this case we find that $\Delta I(\alpha) = \Delta I(\gamma)$, and $\Delta I(\beta) = \Delta I(\delta)$. This is easily explained by rotating the xy-plane with 180°

$$\begin{aligned} \Delta I(\gamma) &= I^{+}_{x-}(\gamma) - I^{-}_{x-}(\delta) \\ &= I^{+}_{x+}(\alpha) - I^{-}_{x+}(\beta) = \Delta I(\alpha). \end{aligned} \tag{10}$$

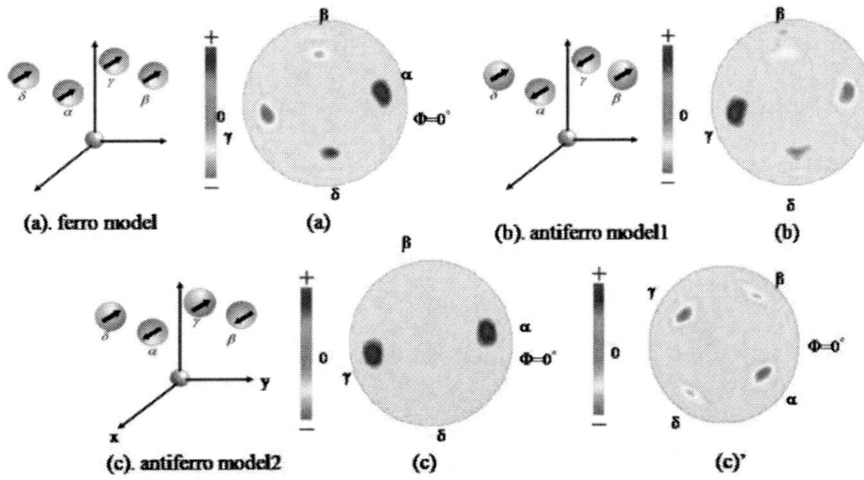

Figure 2. Left: Cu-Gd_4 model magnetized parallel to the surface: (a) ferro, (b) antiferro1 (α and γ are down spin), (c) antiferro2 (α and β are magnetized with down spin). Right: Calculated modified CDAD $\Delta I(\theta, \phi)$. The calculated $\Delta I(\theta, \phi)$ are shown for the model (c) by (c) and (c)', where α and γ are in the xz-plane, and xy-plane is rotated with 45° around z-axis. In (c) we have no mirror symmetry with respect to xz-plane, whereas in (c)' we have the symmetry.

Journal of Physics: Conference Series **235** (2010) 012009 doi:10.1088/1742-6596/235/1/012009

3.3. Fe/Cu ferro model with perpendicular magnetization

So far we have considered photoemission from one Cu atom. In real surface XPD spectra the photoemission from the second, the third, ... layers also contributes to the dichroism. As examples of more realistic systems, we consider iron monolayer and double layers on Cu(001) substrate. The model cluster is shown in Figure 3. Figure 3(a) and (b) show the calculated modified circular dichroism $\Delta I(\theta, \phi)$ for the ferromagnetic model with perpendicular magnetization. In (a) Fe monolayr and (b) Fe double layers on Cu substrate (ϵ_k = 90 eV) are considered. The peak image from double-layer model is more complicated than from the monolayer model. We still have clear imaging of the magnetic surface layers.

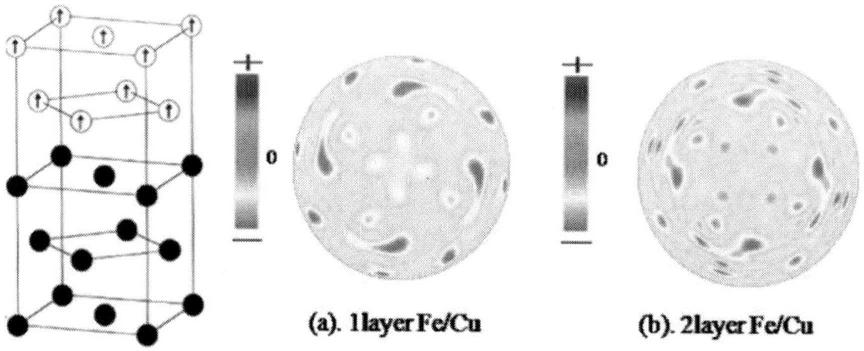

Figure 3. 2 layer Fe/Cu model (left). Calculated modified CDAD $\Delta I(\theta, \phi)$ for (a) 1 layer Fe/Cu ferromagnetic model and (b) 2 layer Fe/Cu ferromagnetic model.

3.4. Fe/Cu ferro model with in plane magnetism

We next discuss the in-plane magnetized model of the same surface systems considered in Figure 3. Figures 4(a) and (b) show the calculated modified $\Delta I(\theta, \phi)$ for the ferromagnetic model magnetized in plane. In (a) Fe monolayer and (b) Fe double-layer on Cu substrate (ϵ_k = 90 eV) are considered. The peak images in (a) and (b) become complicated and it is difficult to obtain useful information on surface magnetic order.

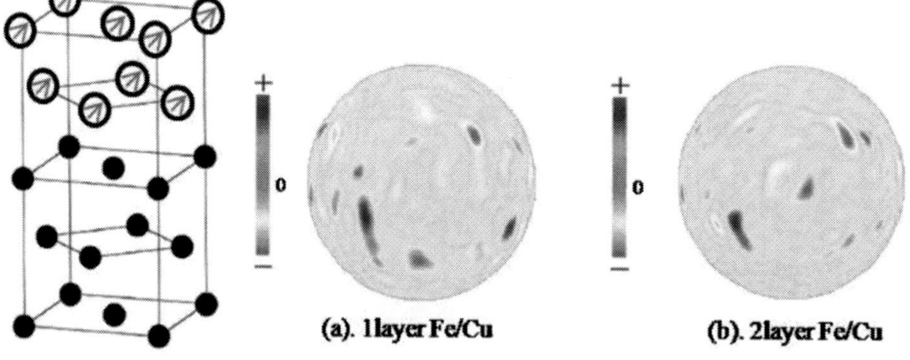

Figure 4. 2 layer Fe/Cu model (left). Calculated modified CDAD $\Delta I(\theta, \phi)$ for (a) 1 layer Fe/Cu ferromagnetic model and (b) 2 layer Fe/Cu ferromagnetic model.

4. Conclusion

We proposed a possible method to obtain only direct imaging of magnetic atoms by use of the photoelectron diffraction method that does not require any expensive spin detector. This approach applies the cancellation of the Daimon effect which rotates the forward focusing PD peaks with finite angles. The modified circular dichroism $\Delta I(\theta, \phi)$ picks up the information only of magnetic atoms. Illustrative numerical calculations clearly distinguish the magnetic atoms with different spin orientation in the case of the single magnetic layer on Cu substrate. Thicker magnetic layer, for example, double magnetic layers make the atomic image unclear because of increasing the number of emitter. The method is useful to study magnetic monolayer on nonmagnetic substrate. The images obtained for in-plane magnetized models show rather complicated features, however their symmetric patterns are well explained by use of symmetry consideration. We hope experimental studies will be performed to obtain the imaging magnetization as we proposed here.

Acknowledgement

The authors are grateful to have fruitful discussions with Dr. H. Shinotsuka.

References

[1] J. Bansmann, C. Ostertag, G. Schönhence, F. Fegel, C. Westphal, M. Getzlall, F. Schafers, H. Petersen, Phys. Rev. B **46** 13496 (1992)
[2] H. Daimon, T. Nakatani, S. Imada, S. Suga, Y. Kagoshima, T. Miyahara, Jpn. J. Appl. Phys. **32** L1480 (1993)
[3] T. Fujikawa, M. Yimagawa, J. Phys. Soc. Jpn. **63** 4220 (1994)
[4] T. Fujikawa, Physica B **93** 208-209 (1995)
[5] P. Rennert, A. Chassé, T. Nakatani, K. Nakatsuji, H. Daimon, S. Suga, J. Phys. Soc. Jpn. **66** 396 (1997)
[6] A. Chassé, P. Rennert, Phys. Rev. B **55** 4120 (1997)
[7] H. Daimon, Phys. Rev. Lett. **86** 2034 (2001)
[8] T. Nakatani, T. Matsushita, Y. Miyake, T. Nohno, A. Kobayashi, K. Fukumoto, S. Okamoto, A. Nakamoto, F. Matsui, K. Hattori, M. Kostugi, Y. Saitoh, S. Suga, H. Daimon, Prog. Surf. Sci. **71** 217 (2003)
[9] F. Matsui, T. Matsushita, F.Z. Guo, H, Daimon, Surf. Rev. Lett. **14** 1 (2007)
[10] M. Wada, K. Ito, T. Konishi, T. Fujikawa, Surface Science **602** 2907-2914 (2008)
[11] H. Shinotsuka, H. Arai, T. Fujikawa, Phys. Rev. B. **77** 085404 (2008)
[12] K. Ito, T. Konishi, T. Fujikawa, J. Surf. Anal. **14** 328 (2008)
[13] K. Ito, T. Konishi, T. Fujikawa, e-J. Surf. Nanotech. **7** 115 (2009)

Journal of Physics: Conference Series **235** (2010) 012010 doi:10.1088/1742-6596/235/1/012010

X-ray Photoelectron Diffraction Study of Bias-Treatment for the Growth of 1-Inch-Diameter Hetero-Epitaxial Diamond (001) Thick Films

S. Kono[1]*, H. Kawata[1], T. Goto[1], T. Abukawa[1], K. Chigira[2,3], K. Ooyama[4], T. Kotaki[4] and A. Sawabe[2,3†]

[1] IMRAM, Tohoku University, Sendai 980-8577, Japan

[2]Electri. & Electron. Engi., Aoyama Gakuin University, Sagamihara 252-5258, Japan

[3]AGD Material Co., Ltd., Sagamihara 252-5258, Japan

[4]Namiki Precision Jewel Co., Ltd., Adachi-ku, Tokyo 123-8511, Japan

*Email: kono@tagen.tohoku.ac.jp. Present affiliation: Electri. & Electron. Engi., Aoyama Gakuin University.

† Email:sawabe@ee.aoyama.ac.jp

Abstract. Bias-treatment (BT) is widely used as an initial seeding process of chemical vapor deposition (CVD) growth of diamonds. In a BT, substrates are negatively biased by an order of hundred volts so that positive ions in the plasma of CVD gas source bombard the substrate to form 'seeds' of diamond growth. Bias-treated 1-inch-diameter (ϕ) Ir(001)/MgO(001) substrates were used as the substrates of the CVD growth of 1-inch- ϕ diamond (001) thick films. XPS (X-ray Photoelectron Spectroscopy) spectra and C 1s and Ir 4f XPD (X-ray Photoelectron Diffraction) patterns were measured for the BT 1-inch- ϕ samples. 5 typical portions on the sample were measured; one at the central portion and the other 4 were peripheral portions, 90° apart from each other and ~4 mm from the edge of the sample. The uniformity of the result over the 5 portions was confirmed. The coverage of carbon layers on the sample after BT was evaluated to be ~ 50 ML in the unit of monolayer of Ir(001)1x1 surface. A quantitative measure of the effect of BT, $A_{45°}$ (the degree of variation of XPD pattern as defined as I_{max}-I_{min}/I_a , where I_{max}, I_{min} and I_a are the maximum, minimum and average intensities of photoelectrons in an azimuthal diffraction pattern at the polar angle of 45°, respectively), was examined for C 1s and Ir 4f core-levels for the 1-inch- ϕ BT samples. The $A_{45°}$ values appeared to be ~3% and ~17% for C 1s and Ir 4f, respectively. The results were compared with those for other bias-treated smaller samples reported so far. The $A_{45°}$ value of ~3% for C 1s is found to be very small compared to the corresponding values for the smaller samples reported so far. Thus, the method of BT for the present 1-inch- ϕ sample was found to have a large margin of improvement in terms of $A_{45°}$ for C 1s XPD.

1. Introduction

Bias-treatment (BT) or bias-enhanced nucleation (BEN) is widely used as an initial seeding process of chemical vapor deposition (CVD) growth of diamonds. In a BT, substrates are negatively biased by an order of hundred volts so that positive ions in the plasma of CVD gas source bombard the substrate to form 'seeds' of diamond growth. A BT on an Ir(001) substrate has been first reported by Ohtsuka et al. [1] to grow hetero-epitaxial diamond (001) films on the Ir(001) substrate. Since then, growth of hetero-epitaxial diamond (001) films on Ir(001) substrates has been reported by many workers [2-8]. So far, the crystalline quality of hetero-epitaxial diamond films grown on Ir(001) substrates has been the best among other hetero-epitaxial diamond films in terms of the peak sharpness of diamond Raman shift and of X-ray diffraction locking curves [9, 10]. Application of the hetero-epitaxial diamond (001) films to devices has just started [11]. For the further application of CVD diamond to devices, hetero-epitaxial diamond films on Ir substrates with a sufficiently large size are desired. Fabrication of large size (typically larger than ~1 inch diameter) diamond films has been just tried [10, 12, 13].

The dependence of crystalline quality of hetero-epitaxial diamond films to the quality of BT has been studied previously [8, 9, 14]. In these studies, X-ray photoelectron diffraction (XPD) patterns of C 1s core-level appeared to be most sensitive to the quality of BT. In particular, the anisotropy, A_{45° (the degree of variation of XPD pattern as defined as I_{max}-I_{min}/I_a , where I_{max}, I_{min} and I_a are the maximum, minimum and average intensities of C 1s photoelectrons in an azimuthal diffraction pattern at the polar angle of 45°, respectively) appeared to be a good measure of the quality of BT. Actually, C1s XPD patterns after BT have been first reported to be similar to the C 1s XPD patterns of single crystal diamond (001) surface [15]. The A_{45° corresponds to the degree of forward scattering peak between the nearest neighbor carbon atoms along <110> direction in nanometer scale diamond crystallites epitaxially grown on the Ir(001) substrate[15]. It was also found that there are non-oriented carbon layers that give no XPD peaks. The volume ratio of epitaxial diamond crystallites to the non-oriented carbon layers gives the variation in A_{45°[8, 15]. It was recently suggested that, in addition to the epitaxial diamond crystallites, there are carbon atoms in a highly defective crystalline structure which can contribution to the XPD pattern [14]. In any cases, the A_{45° is a good measure of the quality of BT.

The A_{45° for a homoepitaxial diamond (001) wafer was ~32% [16] while that was ~67% for a thick hetero-epitxial diamond(001) film[14]. The reason for this large difference is not certain but likely to be the difference in angular resolutions of electron spectrometers; A gray scale stereograph (Fig.4 of Ref.[16]) of C 1s XPD pattern has less fine structures than a corresponding stereograph (Fig.3b of Ref.[14]). A commercial electron spectrometer (CLAM4) was used in the former experiment and another commercial electron spectrometer (VG ESCALAB220) was used for the later. C 1s XPD patterns of a thick hetero-epitaxial diamond (001) films was later measured with the third electron spectrometer (ADES400) and A_{45° appeared to be ~50 %[17]. Therefore, we can state that A_{45° value for a bulk diamond (001) can be a value between ~30% and ~70% depending on the angular resolution of electron spectrometer used.

After a BT using a three-electrode DC plasma apparatus, the A_{45° was ~46% with ADES400 spectrometer [15]. Thus, the BT using the three-electrode DC plasma apparatus gives almost perfect alignment of seeding diamond crystallites. However, the area of BT was limited only to a circle of ~2.5 mm in diameter[15]. The A_{45° value for a BT sample in a micro-wave (MW) plasma CVD chamber was ~12% with ESCALAB220 spectrometer which should be compared to the A_{45° value of ~67% for the thick epitaxial diamond film[14]. The area of BT region for this sample was not strictly described but the substrate size was 10x10mm^2. A BT method on an area of 10x10mm^2 was invented using a DC plasma apparatus [9]. The A_{45° values for the 10x10mm^2 area BT samples was examined with ADES400 to have typical values of 9-22% which should be compared to the A_{45° value of ~50% of the thick epitaxial diamond (001) film [8]. A free standing diamond wafer of a size of ~10x10mm^2 was made from a BT sample similar to the above and the FWHM (full width at half maximum) of (004) X-ray diffraction locking curve was found to be ~0.13° [9]. A one-inch ϕ free standing diamond wafer was made from a one-inch-ϕ BT Ir(001)/MgO(001) substrate and the FWHM of (004) X-ray diffraction locking curve was found to be ~0.35° [12]. Thus, the fabrication method of the one-inch-ϕ diamond (001) wafer has not reached the optimized stages for the 10x10 mm^2 substrates. The purpose of the present study is therefore to examine the goodness or badness of BT of one-inch-ϕ Ir(001)/MgO(001) substrates by measuring $A_{45°}$ value in C 1s XPD patterns.

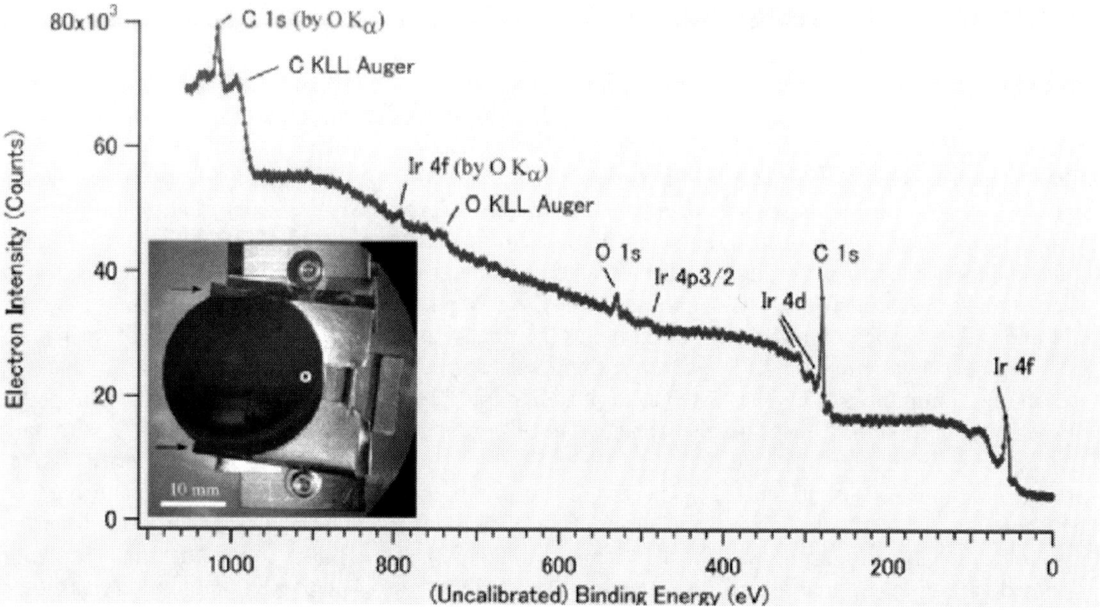

Figure 1. Wide-scan XPS spectrum measured at the photoelectron polar angle of 45° from the central portion of the 1-inch-ϕ BT sample. Inset shows a photograph of the sample holder and 1-inch-ϕ sample (a dark disk).

2. Experimental

The samples used were 1-inch-ϕ Ir(001) films (thickness~500nm) which were epitaxially grown on mirror-polished 1-inch-ϕ MgO(001) substrates. BTs were performed on the substrates in a planar-diode DC-plasma chamber. Typical conditions of BT were: bias voltage-current, ~390V-1.35A; source gas/ratio and pressure, $CH_4/(CH_4+H_2)$=~2% at 120 Torr; duration, 30-45s. These conditions were the best so far in terms of crystalline quality of 1-inch- ϕ diamond (001) films [18]. The BT sample was then contained in a desiccator and transferred to a UHV photoelectron spectrometer (ADES400) which is the same apparatus used previously for the 10x10 mm^2 samples [8].

To facilitate XPS and XPD measurements of the 1-inch-ϕ samples, a special sample holder was made of a molybdenum (Mo) sheet. Inset of Fig.1 shows a photograph of the holder and a 1-inch-ϕ sample (a dark disk). The 1-inch-ϕ sample is loosely pressed by a Mo spring from the back-side outward the two straight stopper edges (as marked by two arrows) at the upper and lower sides. The position and area of XPS and XPD measurements are marked by a white circle on the right hand side of the dark disk, which is the midway between the two screw heads on the up and down sides. For XPD measurement, the whole the sample holder was rotated about an axis located at the center of the white circle and normal to the 1-inch-ϕ disk surface. To change the position of measurements, the sample disk was slid sideway and rotated about the surface normal by a flat tip end of a transfer-rod connected to the outside of vacuum. The sample position could be judged by several position markers made on the backside of the sample disk. Typical positions examined were a) central portion, 4 peripheral portions 90° apart from each other and ~4 mm from the edge of the sample disk.

3. Results and discussion

Figure 1 shows a typical wide-scan XPS spectrum (excited by Mg K_α lines) measured at photoelectron polar angle of 45° from the central portion of the sample; Wide-scan XPS spectra measured for the other 4 peripheral portions showed similar spectra. It must be noted that the mode (a constant pass energy mode) of wide-scan is different from that (a constant retard ratio mode) for Fig.2 of the previous similar study[8], thus the intensity distribution appears differently. The assignment of apparent peaks is given in Fig.1, where C 1s and I 4f peaks are the main components; C1s and I 4f peaks excited by O K_α X-ray from an oxidized Mg target in the X-ray tube are also seen. Impurity components, O 1s XPS and O KLL Auger peaks are also discernible. No other unexpected major impurity components can be seen in Fig.1. The coverage of oxygen can be roughly estimated assuming a uniform oxide overlayers and taking into account of the O1s/Ir4f photo-excitation cross-section ratio and electron attenuation length [19, 20]. The oxygen coverage turned out to be ~ 5 layers in the unit of Ir(001) layer, which is reasonable to be due to the air transfer of the BT sample.

Figure 2 shows a detailed XPS spectrum for C 1s, Ir $4d_{5/2}$ and Ir $4d_{3/2}$ core levels measured under the same condition as for Fig.1. Similar spectra were measured for the other 4 peripheral portions and the peak intensity ratios of C 1s to Ir $4d_{5/2}$ are summarized in Table I. The intensity ratios can be very roughly converted to the coverage of carbon layers assuming a uniform carbon overlayer and taking into account of C1s/Ir$4d_{5/2}$ photo-excitation cross-section ratio and electron attenuation length [19, 20]. The result is summarized in Table I; the coverage is in the unit of monolayer (ML) of diamond (001) layer which is 0.089 nm in thickness.

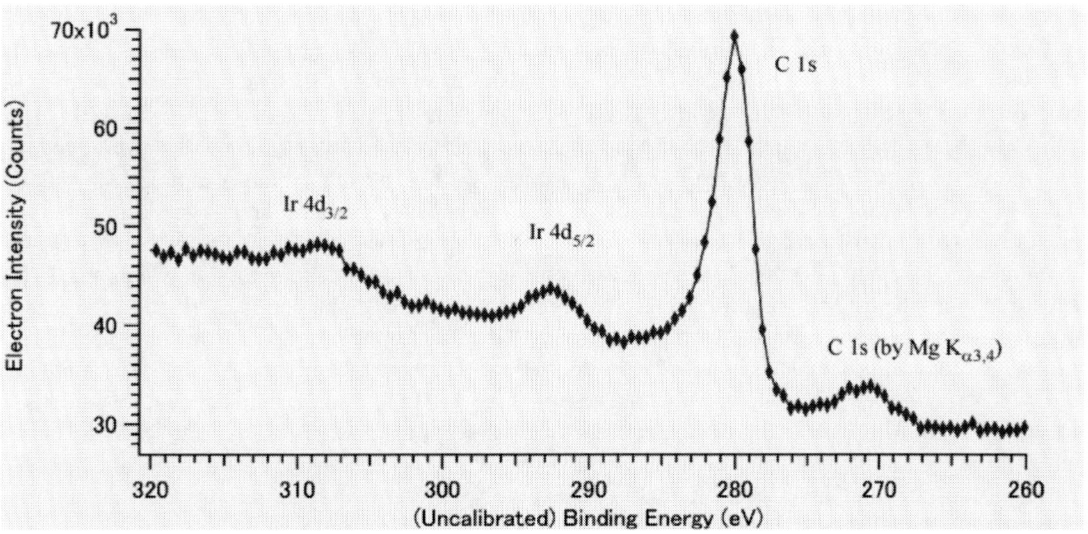

Figure 2. Detailed XPS spectrum for C 1s, Ir $4d_{5/2}$ and Ir $4d_{3/2}$ core levels measured for the 1-inch-ϕ BT sample under the same condition as in Fig.1.

Figure 3 shows azimuth-dependent XPD patterns of C 1s (red circles) and Ir 4f (blue squares) core-levels measured on the central portion of the 1-inch-ϕ BT sample at the polar detection angle of 45°. The abscissa is the azimuthal angle of photoelectron detection and the <100> and <110> axes of substrate Ir(001) surface are marked in Fig.3. The ordinates on the right and left hand sides are the anisotropy of C 1s and Ir 4f XPD patterns, respectively. The anisotropy is defined as $I\text{-}I_a/I_a$, where I is the electron intensity at an azimuthal angle and I_a the average of electron intensity over the azimuthal scan. The $A_{45°}$ value for Ir 4f in Fig.3 is found to be ~17% and that for C 1s in Fig.3 is found to be rather low of ~2.5%. Similar azimuthal XPD patterns were measured for the other 4 peripheral portions and the resulting $A_{45°}$ values are summarized in Table I.

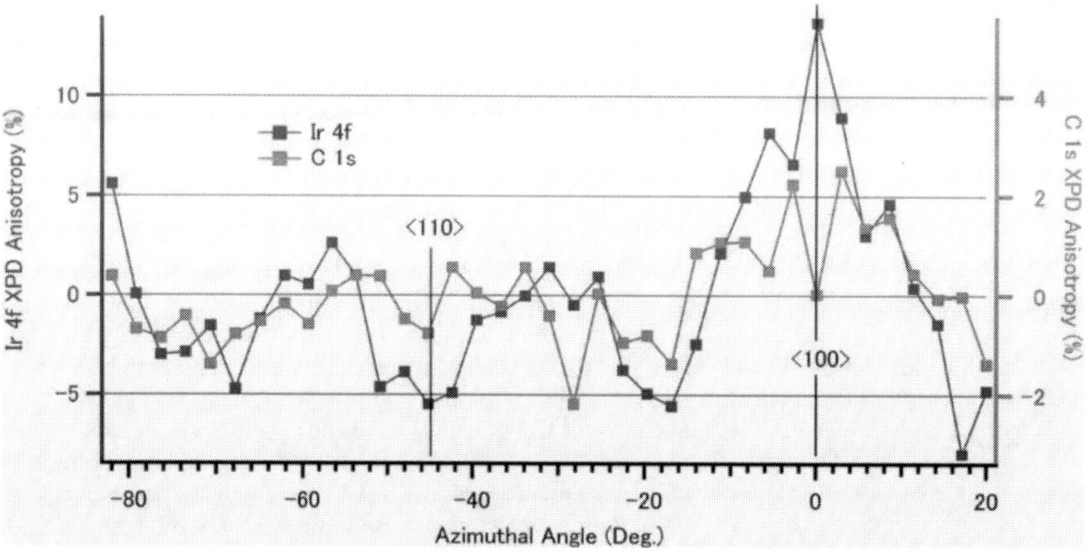

Figure 3. Azimuthal dependent XPD patterns of C 1s (red squares) and Ir 4f (blue squares) core-levels measured on the central portion of the 1-inch-ϕ BT sample at the polar detection angle of 45°.

As found in Figs.1-3 and in Table I, the uniformity of carbon coverages and $A_{45°}$ values is expected to be good over the whole 1-inch-ϕ substrate. However, the mean $A_{45°}$ value for C 1s is ~3 % which is far below the $A_{45°}$ values of ~15-22 % found for 10x10mm^2 substrate [8]. The carbon coverage of ~50 ML for the 1-inch-ϕ substrate is also slightly higher than that of ~36-40 Ml for the 10x10 cm^2 substrate [8]. The fraction of epitaxial nano-diamond (001) crystallites to the whole carbon coverage may be roughly evaluated from the mean $A_{45°}$ values of ~3 % for the 1-inch-ϕ sample using the $A_{45°}$ value of ~50% for the hetero-epitaxial diamond (001) films [15]. For simplicity, we assume that there are only two kinds of carbon layers, one is non-oriented carbon layers and the other is nano-scale diamond (001) epitaxial crystallite layers which are present at the boundary between the Ir(001) substrate and the ~50 ML carbon layer. Reason for the $A_{45°}$ value reduction is ascribed to the fact that non-orientated carbon layer shows no anisotropy. The fraction x of $I_a(epi)$ from epitaxial layer to $I_a(non)$ from non-oriented carbon layer is defined as $I_a(epi)=xI_a(non)$ and x can be evaluated as $0.03=0.5\times xI_a(non)/(xI_a(non)+I_a(non))$ which results that x=~0.064. Taking into account the attenuation length (~2 nm [21]) of C 1s photoelectrons from the epitaxial diamond layers, the thickness of epitaxial layer can be evaluated to be ~1.6 nm (~18 ML). This rather large value indicates that most parts of the epitaxial layers are in reality in a highly defective crystalline structure as suggested previously [14]. The two dimensional density (D) of epitaxial crystallites can be indirectly evaluated from scanning electron microscopic image of a sample subjected to a short-time diamond CVD growth of the BT substrates [9]. The density appeared to be ~10^{10} cm^{-2} for a 10x10mm^2 sample with 30s diamond CVD growth for a BT sample with $A_{45°}$ of ~18% [9]. The density appeared to be ~2x10^9 cm^{-2} for a 10x10mm^2 sample with 10 min diamond CVD growth for a BT sample with $A_{45°}$ of ~18% [8]. The density D appeared to be ~2x10^{8}cm^{-2} for the present 1-inch-ϕ sample with 30 s diamond CVD growth [18]. Thus, the method of BT for the present 1-inch-ϕ sample has a large margin of improvement in terms of $A_{45°}$ and D.

The $A_{45°}$ values for Ir 4f are ~15-18% as found in Table I. The $A_{45°}$ values for Ir 4d$_{5/2}$ was ~25% for a 10x10 mm^2 sample with C 1s $A_{45°}$ of ~18% [17]. It was found that $A_{45°}$ value fro Ir 4f was not much

Journal of Physics: Conference Series **235** (2010) 012010 doi:10.1088/1742-6596/235/1/012010

different from $A_{45°}$ value for Ir $4d_{5/2}$ [17]. Thus, the present 1-inch- ϕ BT sample is not much far from the 10x10 mm^2 BT sample in terms of $A_{45°}$ values for Ir 4f or $4d_{5/2}$.

4. Conclusion

XPS spectra and C 1s and Ir 4f XPD patterns were measured for bias-treated 1-inch-ϕ Ir(001)/MgO(001) samples which are the substrates of the CVD growth of 1-inch-ϕ diamond (001) thick films. 5 typical portions on the sample were measured; one at the central portion and the other 4 are peripheral portions, 90° apart from each other and ~4 mm from the edge of the sample. The XPS and XPD results over the 5 positions showed a good uniformity of the bias-treatment. The coverage of carbon layers on the sample after BT was evaluated to be ~ 50 ML in the unit of monolayer of diamond (001)1x1 surface. A quantitative measure of quality of BT, $A_{45°}$, was examined for C 1s and Ir 4f core-levels for the 1-inch-ϕ BT samples. The $A_{45°}$ values appeared to be ~3% and ~17% for C 1s and Ir 4f, respectively. The results were compared with those for other bias-treated smaller samples reported so far. The $A_{45°}$ value of ~3% for C 1s is found to be very small compared to the corresponding values for the smaller samples reported so far. Thus, the method of BT for the present 1-inch-ϕ sample was found to have a large margin of improvement in terms of $A_{45°}$ for C 1s XPD.

References

[1] K. Ohtsuka, K.Suzuki, A. Sawabe and T. Inuzuka, Jpn. J. Appl. Phys. **35**(1996)L1072.
[2] K. Ohtsuka, H. Fukuda, K. Suzuki and A. Sawabe, Jpn. J. Appl. Phys. **36**(1997)L1214.
[3] M. Schreck, H. Roll, and B. Stritzker, Appl. Phys. Lett. **74**(1999)650.
[4] C. Bednarski, Z. Dai, A.-P. Li, and B. Golding, Diamond Relat. Mater. **12**(2003)241.
[5] Z. Dai, C. Bednarski-Meinke, R. Loloee, and B. Golding, Appl. Phys. Lett. **82**(2003)3847.
[6] S. Gsell, T. Bauer, J. Goldfuß, M. Schreck, and B. Stritzker, Appl. Phys. Lett. **84**(2004)4541.
[7] T. Bauer, S. Gsell, M. Schreck, J. Goldfuß, J. Lettieri, D. G. Schlom, and B. Stritzker, Diamond Relat. Mater. **14**(2005)314.
[8] T. Aoyama, N. Amano, T. Goto, T. Abukawa, S. Kono, Y. Ando, A. Sawabe, Diamond Relat. Mater. **16**(2007)594.
[9] S. Ichihara, Y. Ando, K. Suzuki, N. Amano, S. Kono and A. Sawabe, Extended Abstracts of 20st Diamond Symposium, p52, Tokyo, Nov. 21, 2006 (in Japanese).
[10] M. Fischer, S. Gsell, M. Schreck, R. Brescia, B. Stritzker, Diamond & Rela. Mater. **17**(2008)1035.
[11] Y. Yokota, N. Kawakami, Y. Ando, T. Tachibana, K. Kobashi, A. Sawabe, New Diamond & Frontier Carbon Tech. **17**(2007)2.
[12] K. Okada, K. Chigira, Y. Ando, K. Suzuki and A. Sawabe, Extended Abstracts of 20st Diamond Symposium, p4, Tokyo, Nov. 21, 2006 (in Japanese).
[13] M. Fischer, R. Brescia, S. Gsell, M. Schreck, T. Brugger, T. Greber, J. Osterwalder and B. Stritzker, J. Appl. Phys. **104**(2008)123531.
[14] S. Gsell, S. Berner, T. Brugger, M. Schreck, R. Brescia, M. Fischer, T. Greber, J. Osterwalder, B. Stritzker, Diamond & Rela. Mater. **17**(2008)1029.
[15] S. Kono, T. Takano, T. Goto, Y. Ikejima, M. Shiraishi, T. Abukawa, T. Yamada and A. Sawabe, Diamond Relat. Mater. **13**(2004)2081.
[16] S. kono, T. Takano, M. Shimomura, T. Goto, T. Abukawa, M. Tachiki and H. Kawarada, Surf. Sci. **529**(03)180.
[17] N. Amano, Master Thesis, Tohoku University (2006).
[18] K. Chigira & A. Sawabe, private communications.
[19] J. J. Yeh and I. Lindau, Atomic Data & Nuclear Data Tables **32**(1985)1.
[20] S. Tanuma, C. J. Powell, D. R. Penn, Surf. Interface Anal. **17**(1991)911.
[21] S. Kono, T. Saitou, H. Kawata, T. Goto, Y. Kakefuda, T. Komeda, Surf. Sci. **603**(2009)860.

Excitation power dependence of lifetime of photoluminescence from gold film in the Kretschmann geometry

Tomonori Sanbongi[1], Satoshi Katano[1], Jamal Uddin Ahamed[1], Yoichi Uehara[1,*] and Sukekatsu Ushioda[2]

[1] Research Institute of Electrical Communication, Tohoku University, 2-1-1 Katahira, Aoba, Sendai, Miyagi, 980-8577 Japan

[2] National Institute for Materials Science, 1-2-1 Sengen, Tsukuba, Ibaraki, 305-0047 Japan

Abstract. We have investigated picosecond-laser-pulse-excited photoluminescence (PL) from an evaporated gold film in the Kretschmann geometry. The total intensity of PL was proportional to the square of continuous wave equivalent laser power for excitation, showing that PL is excited via two-photon optical processes. The lifetime of PL was directly determined as a function of excitation laser power by use of a streak camera. We found that the lifetime is inversely proportional to excitation laser power, indicating that the lifetime of d-band holes created by the two-photon optical processes is inversely proportional to their density.

1. Introduction

Intensive studies of photoluminescence (PL) of gold (Au) have been motivated not only by fundamental interests but also by industrial applications, such as plasmonic materials [1]. PL from Au was first reported by Mooradian in 1969 [2]. Boyd et al. measured excitation power dependence of PL of Au films, and investigated single- and multi-photon excitation processes in PL [3]. Although the dynamics in PL has been intensively investigated for Au, reported lifetimes of PL have ranged from several tens of femtoseconds to a few nanoseconds, showing that the underlying mechanism is complex and has not been fully understood yet [4-6].

We reported light emission from the W tip-Au sample gap of a scanning tunneling microscope irradiated with picosecond (ps) laser pulses in previous paper [7, 8]. The emission consisted of two components; the first component is excited by the laser-induced electron tunneling whose duration is identical to that of the incident laser pulse and the second component is directly excited by the laser pulse. The duration of the second component was a few hundred ps. This value is much longer than that of the laser pulse used for the excitation.

The second component disappeared when the tip was retracted from the sample surface by a macroscopic distance, showing that this emission is assisted by the presence of the tip. Its spectrum was successfully reproduced by an electromagnetic theory taking into account the presence of the tip which breaks the translational symmetry in the sample surface. We concluded from the experimental

* To whom correspondence should be addressed: uehara@riec.tohoku.ac.jp (Y. U.)

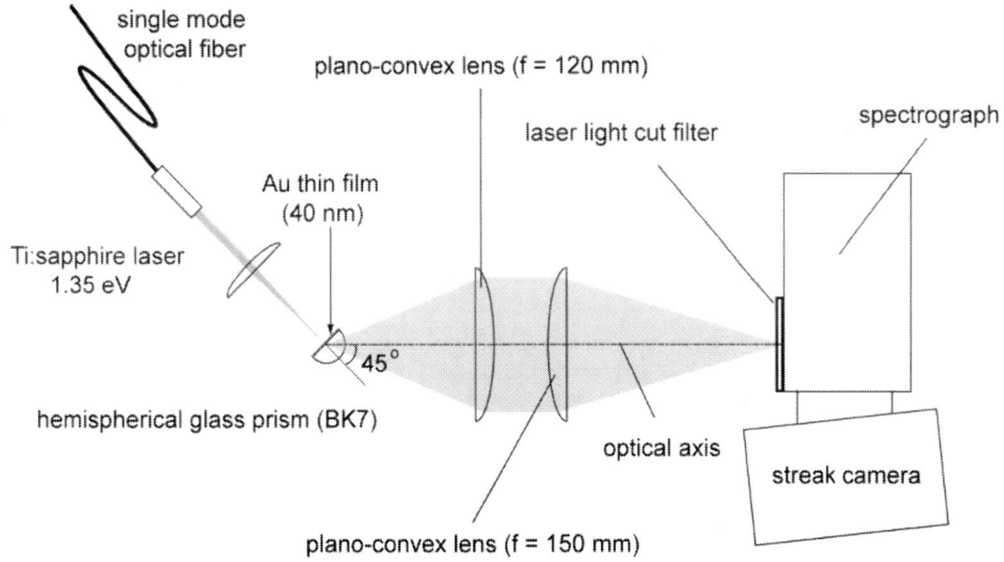

Figure 1. Schematic of experimental setup

and theoretical knowledge that the second component is radiated by surface plasmon polaritons (SPP) through the breakdown of the translational symmetry in the sample surface by the presence of the tip. However, one may notice that the lifetime observed here is quite longer than the lifetime of SPP reported in the past [9].

We have carried out one step further experiment on the second component. In the present experiments, instead of the tip in an STM, the Kretschmann geometry was adopted for making SPP radiative. In this geometry, an Au thin film is deposited on the bottom of a hemispherical glass prism. Since the light propagating in the glass has a wavevector whose magnitude is larger than that of SPP localized at the Au film surface, its wavevector component parallel to the surface can be matched to that of SPP by adjusting the propagation angle of the light. In this manner, SPP becomes radiative with the adjusted angle on the prism side without the tip. We already reported that PL radiated in this adjusted angle has the spectra consistent with those of SPP emission through the prism [10]. In this paper, we report about excitation power dependence of the lifetime of PL.

2. Experimental

The sample was an Au film evaporated on the bottom of a hemispherical glass (BK7) prism in a high vacuum of 5.0×10^{-6} Torr. The mass thickness of the film was 40 nm. The dielectric function of the film measured by ellipsometry was consistent with that of bulk Au [11], showing that the film is continuous. In fact, a continuous grain structure was seen in an atomic force microscope image of the sample. Figure 1 shows the schematic of the experimental setup. Optical measurements were performed in the adjusted angle, which is 45° from the surface normal, under atmospheric conditions. The sample was irradiated with a mode-locked Ti:sapphire laser from the air side to excite PL. The pulse width and the repetition rate of the laser were 2 ps and 80 MHz, respectively. The wavelength of the laser was tuned at the wavelength of 920 nm, which corresponds to a photon energy of 1.35 eV. PL emitted in the direction of 45° from the surface normal through the prism was collimated by a first lens and was focused onto the entrance slit of a spectrograph by a second lens. The spectra were analyzed in the photon energy range from 1.5 to 2.4 eV with the streak camera attached to the spectrograph. The temporal resolution of the streak camera was better than 2 ps in the operation mode used in the present experiments.

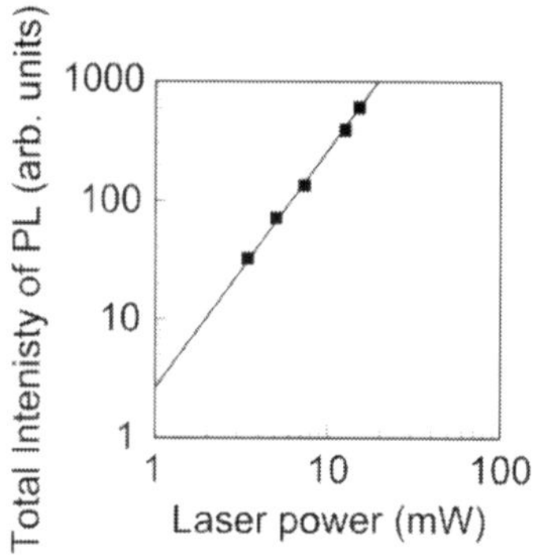

Figure 2. Excitation power dependence of intensity of PL from the Au film.

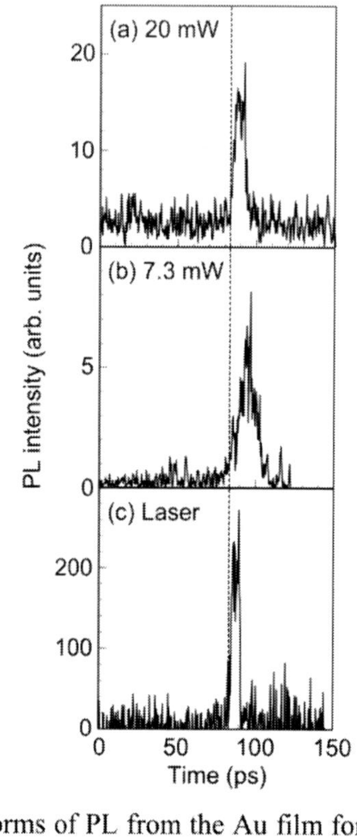

Figure 3. Temporal waveforms of PL from the Au film for excitation laser powers of (a) 20 mW and (b) 7.2 mW, respectively. The temporal profile of the laser pulse is shown in (c).

3. Results and Discussion

Figure 2 shows the continuous wave (cw) equivalent excitation laser power dependence of temporal- and wavelength-integreted (total hereafter) intensity of PL. This graph shows that the total intensity is proportional to the square of excitation laser power. Thus, PL is excited via two-photon optical processes [10].

Figures 3(a) and (b) show temporal waveforms (i.e., wavelength-integrated spectra) of PL for cw equivalent excitation laser powers of 20 mW and 7.3 mW, respectively. Figure 3(c) is the temporal evolution of the laser pulse. The lifetimes of PL read from Figs. 3(a) and (b) are 3 ps and 8 ps, respectively.

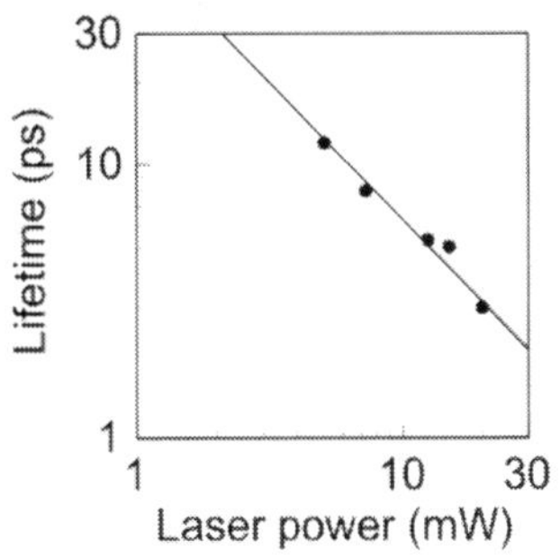

Figure 4. Excitation power dependence of lifetime of PL from the Au film.

The lifetime becomes shorter as the excitation laser power increases. By taking temporal waveforms as shown in Fig. 3 for various values of excitation laser power, we obtained the relationship between the lifetime and the excitation laser power as shown in Fig. 4. From this relation, we see that the lifetime is inversely proportional to the excitation laser power.
As mentioned in the introductory section, various lifetimes have been reported so far. The values reported by Beversluis et al. are in the range of several picoseconds for Au [5]. Those by Imura et al. are from 0.8 ns to 2 ns. Their samples are Au nanorods and are excited by a two-photon optical process. While the lifetime of PL varies in this manner, it is commonly claimed that PL from Au is excited via the recombination of electrons in the sp conduction band and holes created in the d-band by the incident laser. This means that the intensity of PL from Au must be proportional to the number of holes created. In other word, Fig. 4 shows that the lifetime of the holes decreases with the increase of their density. This dependence implies that the density of holes affects their relaxation process through thermally increased electron-electron and electron-phonon scatterings. A decrease of lifetime of PL with an increase of excitation laser power is observed for semiconductors [12-14], which is similar to that shown in Fig. 4. However, the decrease of the lifetime observed for the semiconductors is caused by the screening of internal electric fields which is accelerated by an increase of incident laser power. Thus, this mechanism cannot directly be applied to the present (i.e., metallic substrate) case.

4. Conclusions

Excitation power dependence of PL from Au film with the thickness of 40 nm was investigated in the Kretschmann geometry. Owing to this geometry, the emission from SPP was selectively observed. The total intensity of the emission was proportional to the square of excitation laser power, showing that the light is excited via the two-photon optical processes. Based on the observed power dependence of the lifetime of the emission, we conclude that the lifetime of the d-band holes created by the two-photon optical processes is inversely proportional to its density.

References

[1] Toshima N and Yonezawa T 1998 *New J. Chem.* **22** 1179.
[2] Mooradian A 1969 *Phys. Rev. Lett.* **22** 185.
[3] Boyd G T, Yu Z H and Shen Y R 1986 *Phys. Rev.* B **33** 7923.
[4] Varnavski O P, Mohamed M B, El-Sayed M A and Goodson T *J. Phys. Chem.* B **107** 3101.
[5] Beversluis M R, Bouhelier A and Novotny L 2003 *Phys. Rev.* B **68** 115433.
[6] Imura K, Nagahara T and Okamoto H 2005 J. Phys. Chem. B **109** 13214.
[7] Uehara Y, Yamagami A, Ito K J and Ushioda S 2000 *Appl. Phys. Lett.* **76** 2487.
[8] Uehara Y, Yamagami A, Ito K J and Ushioda S 2001 *Appl. Surf. Sci.* **169-170** 198.
[9] Steinmüller-Nethl D, Hopfel R A, Gornik E and Leitner A 1992 *Phys. Rev. Lett.* **68** 389.

[10] Sanbongi T, Katano S, Uehara Y and Ushioda S, 2008 *Jpn. J. Appl. Phys.* **47** 6114.
[11] Johnson P B and and Christy R W 1974 *Phys. Rev.* B **6** 4370.
[12] Kuroda T and Tackeuchi A 2000 *Appl. Phys. Lett.* **76** 3753-55.
[13] Lee K H, Birner S, Na J H, Taylor R A, Yi S N, Park Y S, Park C M and Kang T W 2006 *Nanotechnology* **17** 5754.
[14] Sari E, Nizamoglu S, Lee I H, Baek J H and Demir H V 2009 *Appl. Phys. Lett.* **94** 211107.

Collision-energy-resolved metastable impact electron spectroscopy of thiophene and 2,2'-bithiophene : Stereodynamics in gas phase and adsorbed layer on Ag(110)

S. Kudo[1], **N. Kishimoto**[1] **and K. Ohno**[2]

[1]Department of Chemistry, Graduate School of Science, Tohoku University, Sendai 980-8578, Japan
[2]Toyota Physics and Chemical Research Institute, Nagakute, Aichi 480-1192, Japan

E-mail: kudosho@mail.tains.tohoku.ac.jp , n-kishi@m.tains.tohoku.ac.jp

Abstract. 2,2'-bithiophene in the gas phase was studied by ultraviolet photoelectron spectroscopy (UPS) and two-dimensional Penning ionization electron spectroscopy (2D-PIES) to obtain the information of stereodynamics in Penning ionization and anisotropic interaction around the ionization region. Thiophene and the 2,2'-bithiophene layer on a Ag(110) surface were investigated by two-dimensional metastable impact electron spectroscopy (2D-MIES). From the relative intensity in the UPS and PIES spectra and the collision energy dependence of partial ionization cross sections (CEDPICS) for 2,2'-bithiophene, ionization bands were assigned to π orbitals. A comparison with the CEDPICS for a gas molecule and the molecular layer on Ag(110) revealed that the attractive interaction between $He^{*}(2^{3}S)$ and both thiophene and 2,2-bithiophene was weak due to the formed molecular layer.

1. Introduction

Investigation of the surface state has become more important for a number of fields interested in nanoscience. Polythiophene and substituted polythiophenes are well-known conductive polymers, which are expected to be good organic electronic devices for molecular level technology [1, 2]. Molecular conductivity depends on the HOMO-LUMO energy gap, and this value is determined by factors such as the polymers substituents, the torsion angle, and the length of conjugation. Thiophene and 2,2'-bithiophene, the monomer and dimer of polythiophene, are the two most basic compounds which can be used to characterize the differences in materials based on factors such as conjugation length and torsion angle. Herein, we applied collision-energy-resolved metastable impact electron spectroscopy to both thiophene and 2,2'-bithiophene in a gas molecule or on a Ag(110) surface, and attempted to compare the anisotropic interaction between $He^{*}(2^{3}S)$ and the thiophene with that between $He^{*}(2^{3}S)$ and the 2,2'-bithiophene.

Electron spectroscopy using excited atoms as the ionization source is known as an effective method to analyze the surface state [3]. This is primarily because the approachable distance between the exited metastable atom and molecule is limited, and selective ionization occurs from the expanded molecular orbitals outside the molecule or the outermost molecular layer. This technique is referred to as Penning ionization electron spectroscopy (PIES) or metastable impact electron spectroscopy (MIES), and was developed by Čermák [4]. The Penning ionization (PI) process, found by Penning [5], is shown in the equation below,

$$M + A^* \rightarrow M^+ + A + e^-. \tag{1}$$

The PI process only occurs in the case of collision between an isolated molecule and a metastable atom. However, when a metastable atom collides with molecule on a metal surface, an electron is emitted by another ionization processes; resonance ionization and Auger neutralization [6]. Therefore, gas phase analysis is referred to as PIES, and surface analysis as MIES.

A coupled technique for selecting the velocity of A^* and analyzing the kinetic energy of the ejected electron has been developed [7] in which the two parameters can be measured simultaneously [8]. The two-dimensional PIES (2D-PIES) or 2D-MIES technique enables one to obtain the collision energy (translated from velocity) dependence of partial ionization cross sections (CEDPICS), thus providing valuable information on anisotropic interactions between metastable atoms and target molecules. This technique has been used to characterize a number of molecules in the gas phase [9]. Recently, we developed an experimental apparatus for surface measurements, and applied measurements to the organic molecular layer.

2. Experimental

The experimental setup for this investigation in the gas phase was previously reported [7, 8]. A metastable He beam was produced by a nozzle discharge source with a tantalum hollow cathode. $He^*(2^1S)$ was quenched using a water-cooled He dc lamp, and pure $He^*(2^3S)$ (19.820 eV) was used as the ionization source for PIES. For ultraviolet photoelectron spectroscopy (UPS), the He I resonance line (21.22 eV) was produced by discharging pure He gas. The emitted electron energy E_e was analyzed using a hemispherical energy analyzer. Calibration of the electron energy scale was made by reference to the lowest ionic state of N_2 mixed with the sample molecule. The background pressure in the chamber containing a collision cell was about 1.0×10^{-4} Pa. He^* was pulsed using a rotating mechanical chopper disk, and time-resolved Penning ionization spectra $I_e(E_e,t)$ were obtained as functions of the electron energy E_e and time-of-flight t. Time-of-flight for $He^*(2^3S)$ was converted to collision energy E_c, and finally the CEDPICS $\sigma(E_e, E_c)$ was obtained. The sample cell and sample line leading to the collision cell were heated to in order to vaporize the sample.

The apparatus for 2D-MIES is shown in Figure 1. The He discharging source and mechanical chopper

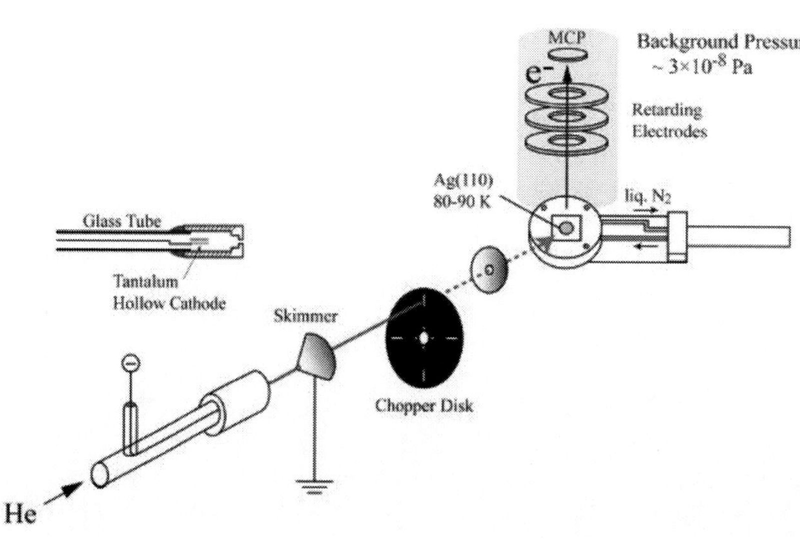

Figure 1. Experimental setup for MIES of a surface.

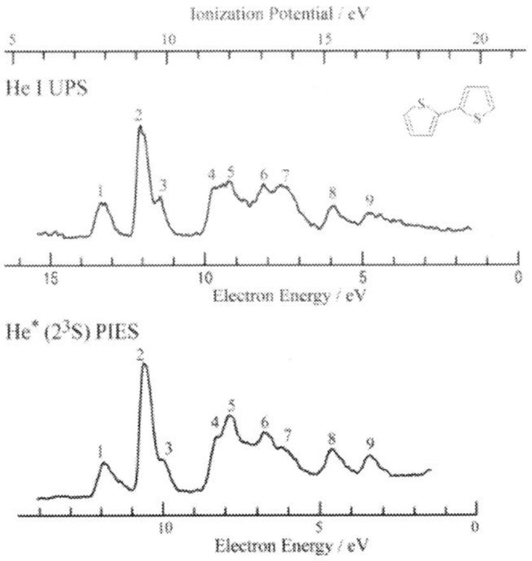

Figure 2. UPS and PIES spectra for 2,2'-bithiophene in the gas phase.

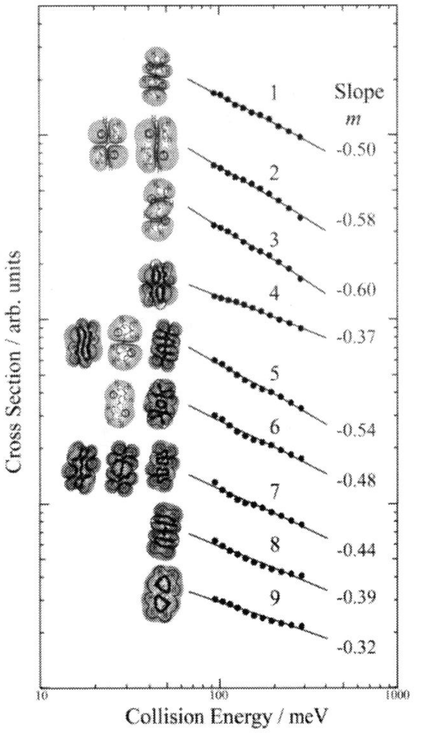

Figure 3. CEDPICS and electron density maps for 2,2'-bithiophene in the gas phase.

for time-resolved measurements were the same as the setup for the gas phase. The electrons were detected by a dual microchannel plate (MCP) at the end of the flight tube through the retarding electrodes placed in the tube [10]. The electron energy spectra were not calibrated. Ag(110) was used as the metal substrate, and it was cleaned by annealing at 800 K and sputtered with Ar^+ prior to experiments. Surface purity was confirmed by Auger electron spectroscopy. The background pressure in the collision chamber was about 3.0×10^{-8} Pa. The sample organic layer was evaporated on a Ag(110) surface after freeze-pump-thaw cycles, and the exposure was determined as a Langmuir unit.

3. Calculations

Quantum chemistry calculations using the Gaussian 03 package program [11] were performed in order to obtain electron density maps, ionization potentials and model potential energy curves. A standard 6-311++G** basis set was used for all calculations. The geometry of thiophene was selected from a microwave spectroscopic study [12], and 2,2'-bithiophene was selected from the optimized structure using density functional theory with the Becke's three-parameter Lee-Yang-Parr functional (B3LYP) [13]. Ionization potentials were calculated by Outer Valence Green's Function (OVGF) method [14]. In order to obtain information on the interaction between He* and the target molecule, potential energy curves in various directions are needed. Since the calculation for highly exited states was difficult, instead of He* a Li atom was introduced in interaction with thiophene or 2,2'-bithiophene because of the well-known similarity between He*(2^3S) and Li(2^2S) [15]. This calculation was performed using B3LYP.

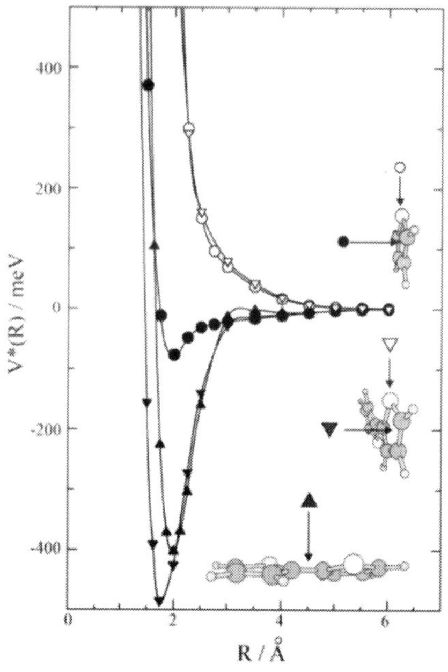

Figure 4. Interaction potential curves V^*(R) for thiophene or 2,2'-bithiophene and Li along various directional approaches. For thiophene-Li, out-of-plane access to the center of the thiophene ring is shown by ● and the S atom from the in-plane direction is indicated by ○. For 2,2'-bithiophene, the center of C2-C2' single bond is indicated by ▲, the center of one thiophene ring from the out-of-plane direction is indicated by ▼, and the S atom from the in-plane direction is indicated by ▽.

4. Results and Discussion

4.1. gas phase analysis by 2D-PIES

Figure 2 shows UPS and PIES spectra for 2,2'-bithiophene in the gas phase. The electron energy scales for PIES are shifted to 1.40 (= 21.22 - 19.82) eV higher than that for UPS by the difference in the excitation energies. Figure 3 shows the log σ vs. log E_c plot of CEDPICS for 2,2'-bithiophene corresponding to each band in Figure 2. The electron density maps, simplified diagrams indicating component atomic orbitals and slope parameter *m* was calculated by the least squares method and are also shown in the figure. Table 1 shows the observed and calculated ionization potential, orbital character and *m* for 2,2'-bithiophene. Results for thiophene [16] are also shown in this table.

Nine clear bands are observed from the UPS and PIES spectra for 2,2'-bithiophene in Figure 2. The ionization potential calculated using the OVGF method shows that some ionization bands overlap in bands 2, 5, 6 and 7. Band 2 was assigned to the π_4 and π_5 orbitals, which is derived from the π_2 orbital of thiophene. The shape of spectrum from band 1 to 3 is very similar to that of biphenyl (C_6H_5-C_6H_5) [17]. When two degenerate $1e_{1g}(\pi)$ orbitals in benzene are split by forming biphenyl, the π_3 and π_6 orbitals of biphenyl with a large band gap and π_4 and π_5 orbitals with a narrow band gap are formed. This suggests that 2,2'-bithiophene and biphenyl have similar electronic structures. The relative intensity of bands 5 and 6 in the PIES spectrum is larger than that in the UPS spectrum. This is because the π_1 and π_2 orbitals, which are expanded out of two thiophene rings, are involved in these two bands, and are advantageous for collision ionization between He* atoms. The relative intensity of band 2 in the PIES spectrum is also larger. On the other hand, that of bands 1 and 3 are not so large because π_6 (band 1) and π_4 (band 4) orbitals have more nodal planes than the other π orbitals. Ionization probability by PI process depends on an overlap between 1s orbital of He* and each molecular orbital, and is lowered around nodal plane. The same trend is observed in the PIES spectra for biphenyl and other aromatic compounds [18]. In addition,

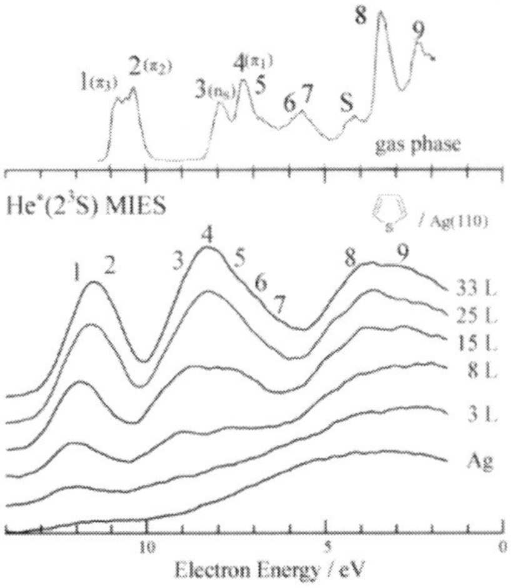

Figure 5. Coverage dependence of MIES spectra for thiophene on Ag(110) surface. The top shows PIES spectrum in the gas phase [16].

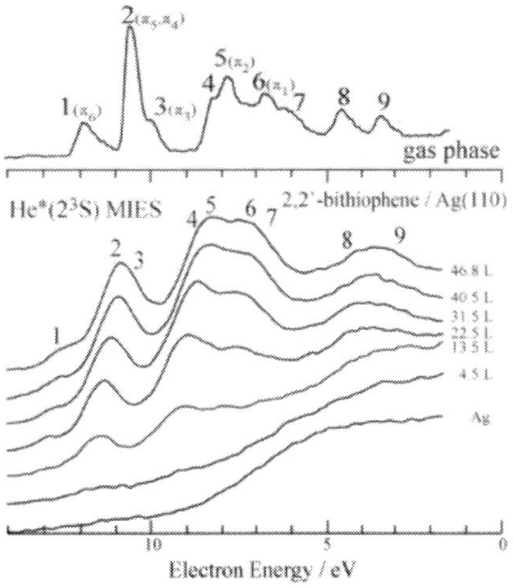

Figure 6. Coverage dependence of MIES spectra for 2,2'-bithiophene on Ag(110) surface. The top shows PIES spectrum in the gas phase.

Figure 3 shows that the CEDPICS for bands 5 and 6 have a strong negative slope (m = -0.54 and -0.48). This is a typical feature for the π orbital region in organic compounds including heterocyclic aromatic compounds [16, 19]. Table 1 shows the CEDPICS for thiophene, which also has a strong negative slope for each band corresponding to a π orbital [16].

The slopes of CEDPICS are quite different for thiophene and 2,2'-bithiophene. The slope of CEDPICS for thiophene corresponding to π bands is -0.26 for band 1, -0.40 for band 2 and -0.33 for band 4. In the case of 2,2'-bithiophene, CEDPICS corresponding to π band 1, 2, 3, 5 and 6 are -0.48 $\sim$ -0.60. This suggests that the attractive interaction for the out-of-plane direction for 2,2'-bithiophene is stronger than that for thiophene, due to the two connecting thiophene rings and the longer conjugated chains. The CEDPICS for the π bands (m = -0.34 $\sim$ -0.45) of biphenyl [17] also show more negative slopes than those of benzene (m = -0.32 $\sim$ -0.40) [19]. On the other hand, the CEDPICS corresponding to band 1 (m = -0.50 for 2,2'-bithiophene and -0.34 for biphenyl) are less negative than band 2 (m = -0.58 for 2,2'-bithiophene and -0.45 for biphenyl) and 3 (m = -0.60 for 2,2'-bithiophene and -0.42 for biphenyl). This results from the fact that π_6 orbital has three nodal planes, and does not disturb the out-of-plane region.

The model potential curves shown in Figure 4 show that the well depth of the interaction potential between Li and one thiophene ring of 2,2'-bithiophene from the out-of-plane direction is about -485 meV, six times deeper than that between Li and thiophene (about -76 meV). From this, theoretical evidence of strong attractive interaction between He* and 2,2'-bithiophene was obtained. The well depth of the interaction potential between Li and the center of the C-C single bond between two thiophene rings for 2,2'-bithiophene is about -400 meV, which has a low value and indicates an expanding π orbital region at the molecular surface due to the connection of the two thiophene rings. Although the model potential shows a repulsive interaction between the Li and S atom from the in-plane-direction for both molecules, n_S or σ bands as well as π bands have a strong negative slope ($m \leq 0$ for thiophene, $m \leq -0.30$ for 2,2'-bithiophene). This is thought to be because π orbital expansion affects He* causing it to approach the n_S or σ orbitals.

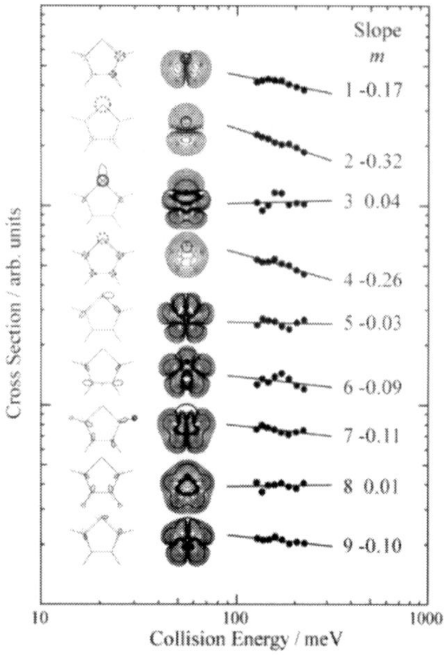

Figure 7. CEDPICS and electron density maps with simplified diagrams indicating component atomic orbitals for thiophene on Ag(110) surface at 25 L.

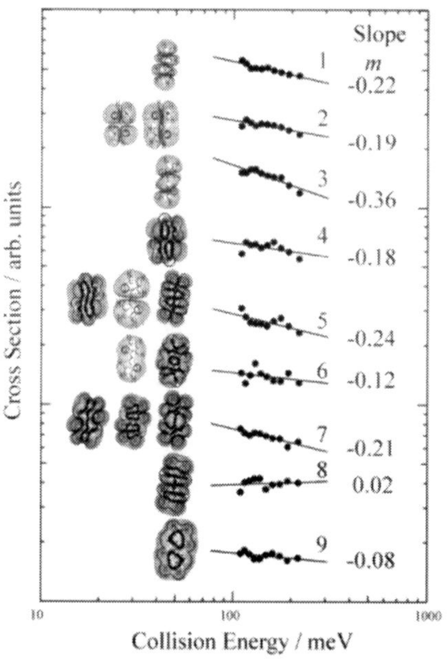

Figure 8. CEDPICS and electron density maps for 2,2'-bithiophene on Ag(110) surface at 23 L.

4.2. surface analysis by 2D-MIES

Figure 5 and 6 show coverage dependence of MIES spectra for thiophene and 2,2'-bithiophene on Ag(110). The lower spectrum is that of a pure Ag(110) surface, and the upper set of spectra are measured during the sample exposure. The top spectrum is that of a gas sample PIES spectrum. Some of the bands observed in the PIES spectrum cannot be seen clearly in the MIES spectra. However, the E_e from PIES spectrum was effectively determined and the IP could be calculated. Figures 7 and 8 show the CEDPICS and electron density maps corresponding to each band in figures 5 and 6. The exposure is 25 L for the thiophene layer and 23 L for the 2,2'-bithiophene layer because each band is sufficiently intense to analyze at these exposures.

The band intensity of thiophene is proportional to the exposure, and the relative intensity of each band is similar to that of the PIES spectrum. This indicates that the outermost molecular orientation is independent of layer thickness and is advantageous in that it ionizes from the π orbital. On the other hand, the intensity of bands 4 and 5 for 2,2'-bithiophene are slightly stronger than those of bands 6 and 7 at lower coverage ($\sim$ 22.5 L). However, the four bands have equal intensity at 46.8 L. Bands 4 and 5 are thought to primarily correspond to the n_S and π orbital and bands 6 and 7 mainly corresponded to the σ orbital, therefore, the band intensity reflects the out-of-plane orbital for bands 4 and 5, and the in-plane orbitals for bands 6 and 7. This spectral change suggests that the in-plane orbital was advantageous in that it ionizes by increasing exposure and molecular inclining from a flat-lying orientation on the substrate. As the layer gets thicker, more spectral change will be observed because interaction between outermost layer and substrate gets weaker and molecular layer forms amorphous or crystal structure. Research on a thiophene and 2,2'-bithiophene monolayer on Ag(111) deposited by NEXAFS [20], suggested that the thiophene ring of both molecules is parallel to the substrate. Furthermore, MIES for thiophene and 2,2'-bithiophene at a lower exposure have the same features.

The negative slope of CEDPICS corresponding to the π orbitals for the thiophene layer on Ag(110)

Table 1. Band assignments, observed and calculated ionization potential(IP) and obtained slope parameter (m) for 2,2'-bithiophene and thiophene.

a) 2,2'-bithiophene

band	IP / eV obs.	IP / eV calc.	orbital character	m gas	m layer
1	7.89	7.72(0.89)[a]	20b(π_6)	-0.50	-0.22
2	9.21	8.95(0.89)	22a(π_5)	-0.58	-0.19
		9.06(0.89)	19b(π_4)		
3	9.71	9.54(0.88)	21a(π_3)	-0.60	-0.36
4	11.53	11.60(0.89)	20a(n_{S2})	-0.37	-0.18
5	11.91	12.01(0.89)	18b(n_{S1})	-0.54	-0.24
		12.38(0.83)	17b(π_2)		
		12.57(0.88)	19a		
6	13.05	13.11(0.85)	16b(π_1)	-0.48	-0.12
		13.28(0.85)	18a		
7	13.57	13.56(0.89)	17a	-0.44	-0.21
		13.84(0.88)	16a		
		13.89(0.88)	15b		
8	15.21	15.64(0.86)	13b	-0.39	0.02
9	16.41	17.12(0.86)	14a	-0.32	-0.08

b) thiophene

band	IP / eV[b] obs.	IP / eV[b] calc.	orbital character	m gas[b]	m layer
1	8.96	8.77	$1a_2(\pi_3)$	-0.26	-0.17
2	9.58	8.98	$2b_1(\pi_2)$	-0.40	-0.32
3	12.04	12.01	$6a_1(n_S)$	-0.15	0.09
4	12.49	12.48	$1b_1(\pi_1)$	-0.33	-0.26
5	13.15	13.29	$4b_2$	-0.04	-0.03
6	13.71	13.44	$5a_1$	-0.05	-0.09
7	14.26	14.52	$3b_2$	-0.13	-0.11
8	16.52	16.88	$4a_1$	-0.19	0.01
9	17.62	18.06	$3a_1$	-0.08	-0.10

[a] Pole strength is shown in parenthesis.
[b] Reference [16]

was observed at bands 1 (m = -0.17), 2 (m = -0.32) and 4 (m = -0.26). These three slopes are less negative than those in the gas phase (-0.26, -0.40 and -0.33, see Table 1), but remain negative compared to other n_S or σ bands (m = 0.09 ~ -0.11). The same trend was observed for CEDPICS corresponding to most π orbitals for the 2,2'-bithiophene layer. Therefore, 2D-MIES to the molecular layer was demonstrated to behave similar to 2D-PIES for gas molecules. On the other hand, these results show an attractive interaction between either thiophene or 2,2'-bithiophene and He*. The out-of-plane interaction is weak when these molecules form molecular layer from the gas phase. This may be due to limiting molecular orientation and rotation. In the gas 2D-PIES study for benzene, increasing slope value m was observed by using supersonic molecular beam comparing to effusive molecular beam [21]. This is because supersonic beam is composed of oriented molecules and probability that the molecular plane of benzene is parallel to the He* beam is decreased. Similar stereodynamic effect can be suggested to the thiophene or 2,2'-bithiophene thin layer of which most thiophene rings are parallel to the surface. For the clarification, more investigation into the mechanism of CEDPICS for the molecular layer should be carried out.

5. Summary

Using UPS, 2D-PIES and 2D-MIES, the gas or molecular layer of thiophene and 2,2'-bithiophene were investigated to study stereodynamics and anisotropic interaction for these molecules. A number of bands in the spectrum for thiophene and 2,2'-bithiophene were assigned to the π orbital based on the UPS and PIES spectra and the negative slope of the CEDPICS. The slope of CEDPICS corresponding to π band for 2,2'-bithiophene in the gas phase was more negative than that for thiophene in the gas phase. This was attributed to the connection between the two thiophene rings and the longer conjugated chain. The difference in the CEDPICS between the gas for thiophene or 2,2'-bithiophene and the molecular layer on the Ag(110) surface was observed. Results suggested that the molecular orientation caused the interaction between 2,2'-bithiophene and He* to be weak.

6. Acknowledgment

This research was supported by a Grant for Scientific Research from the Japan Society for the Promotion of Science (JSPS) and by Matsuo Foundation.

References

[1] Roncali J 1997 *Chem. Rev.* **97** 173
[2] Reddinger J L and Reynolds J R 1999 *Adv. Polym. Sci.* **145** 57
[3] Kubota H, Munakata T, Hirooka T, Kondow T, Kuchitsu K, Ohno K and Harada Y 1984 *Chem. Phys.* **87** 399
[4] Čermák V 1966 *J. Chem. Phys.* **44** 3781
[5] Penning F M 1927 *Naturwissenshaften* **15** 818
[6] Hagstrum H D 1979 *Phys. Rev. Lett.* **43** 1050
[7] Mitsuke K, Takami T and Ohno K 1989 *J. Chem. Phys* **91** 1618
[8] Ohno K, Yamakado H, Ogawa T and Yamata T 1996 *J. Chem. Phys.* **105** 7536
[9] Kishimoto N and Ohno K 2007 *Int. Rev. Phys. Chem.* **26** 938
[10] Yamakita Y, Tanaka H, Maruyama R, Yamakado H, Misaizu F and Ohno K 2000 *Rev. Sci. Instrum.* **71** 3042
[11] Frisch M J et al. *GAUSSIAN 03* Gaussian Inc. Pittsburg, PA, 2003
[12] Nygaard L, Nielsen T, et al. 1969 *J. Mol. Structure* **3** 491
[13] Becke A D 1993 *J. Chem. Phys.* **7** 5648
[14] Ortiz J V 1996 *J. Chem. Phys.* **104** 7599
[15] Niehaus A, 1981 *Adv. Chem. Phys.* **45** 399
[16] Kishimoto N, Yamakado H and Ohno K 1996 *J. Phys. Chem.* **100** 8204
[17] Kishimoto N, Hagihara Y, Ohno K, Knippenberg S, François J P and Deleuze M 2005 *J. Phys. Chem. A* **109** 10535
[18] Ohno K, Mutoh H, and Harada Y, 1983 *J. Am. Chem. Soc.* **105** 4555
[19] Takami T and Ohno K 1992 *J. Chem. Phys.* **95** 6523
[20] Väterlein P, Schmelzer Z, Taborski J, Viczian F, Bäβler M, Fink R, Umbach E and Wurth W 2000 *Surf. Sci.* **452** 20
[21] Horio T, Maruyama R, Kishimoto N, and Ohno K, 2004 *Chem. Phys. Lett.* **383** 73

(e, 2e) processes on Ne, Ar and Xe targets

G. Purohit, Vinod Patidar and K. K. Sud

Department of Physics, School of Engineering, Sir Padampat Singhania University, Bhatewar, Udaipur-313 601, India

g_vpurohit@yahoo.com, ghanshyam.purohit@spsu.ac.in

Abstract. Recently, there have been several attempts to explain the features of triple differential cross section (TDCS) for the (e, 2e) processes on inert targets Ne, Ar and Xe but there are still certain discrepancies in theoretical results and measurements, which require more theoretical efforts to understand the collision dynamics of these targets. We present in this paper the results of our modified distorted wave Born approximation (DWBA) calculation of TDCS for the ionization of Ne (2p), Ar (3p) and Xe (5p) targets. We modify the standard DWBA formalism by including the correlation-polarization potential (which is function of electron density) and compare our computed results with the available experimental and theoretical data. We observe that the polarization potential is able to improve the agreement with experimental results.

1. Introduction

Kinematically complete studies of electron-impact single ionization (i.e. (e, 2e) processes) of atoms and ions provide important information about target structure, target wavefunction and collision dynamics. Since the experiment of Ehrhardt et al. [1] extensive studies have been performed for the electron impact single ionization of hydrogen [2] and helium [3] atoms in different geometrical arrangements. Apart from hydrogen and helium atoms, (e, 2e) studies have been performed on a number of targets using different theoretical approaches (see review articles [4-7]). The study of (e, 2e) processes on heavier atoms, particularly noble gas atoms, has been a subject of recent interest as it introduces physical effects like exchange between active and passive electrons, target polarization etc. which are not evident from the ionization of lighter atoms. There have been numerous theoretical and experimental efforts to explain the trends of triple differential cross section (TDCS) of (e, 2e) processes on Ne, Ar and Xe targets (Waterhouse et al. (1998) [8], Haynes and Lohmann (2000, 2001a, 2001b) [9-11], Biava et al. 2002 [12], Haynes et al. (2003) [13], Stevenson et al. (2005, 2009) [14-15] and Stevenson and Lohmann (2006, 2008) [16,17]. Various theoretical studies have been done to describe the features of the measurement of TDCS in the above mentioned studies. Purohit et al. [18] reported the results of modified distortd wave Born approximation (DWBA) for the ionization from 3s shell of argon atom. Prideaux and Madison [19] have considered the treatment of exchange and post collision interaction (PCI) in the DWBA approach. The hybrid DWBA-R matrix (hereafter referred as DWB2-RM) has been applied to 3p and 3s shell ionization of argon [20], 2p shell ionization of neon [15] and 5p shell ionization of xenon target [15]. DWBA approach which included PCI by a simple Gamow factor (DWBA-G) has also been reported [15, 21]. The distorted wave and 3DW (three distorted wave) approaches [22] reported the results of (e, 2e) processes on Ar(3p). Recently Otranto (2009) [23] reported the results of TDCS for the ionization from the 3p shell of argon atom. All the above mentioned theoretical studies have mixed degree of success in explaining the experimental trends of TDCS for Ne, Ar and Xe targets.

We report the results of our modified distorted wave Born approximation calculations (DWBACPE) for Ne (2p), Ar (3p) and Xe (5p) targets. We modify the standard DWBA formalism by inclusion of correlation-polarization potential as a function of electron density. We present the results of our DWBACPE calculations at incident electron energy 150 eV for Ne (2p) and Xe (5p) atoms and at incident electron energy 113.5 eV for Ar (3p) target. We compare the results of TDCS with the recent available measurements of Stevenson et al (2009) [15] for Ne (2p), Xe (5p) and Stevenson and Lohmann (2008) for Ar (3p) [17]. We also compare the results of our DWBACPE calculations with the recently available theoretical studies [15, 17].

2. Theory

The triple differential cross section for (e, 2e) process on a target may be written as

$$\frac{d^3\sigma}{d\Omega_1\, d\Omega_2\, dE_1} = (2\pi)^4 \frac{k_1 k_2}{k_0} \sum_{av} \left|T(k_1, k_2, k_0)\right|^2 \tag{1}$$

with

$$T(k_1, k_2, k_0) = \left\langle k_1 k_2 \left| T \right| \psi_{nl} k_0 \right\rangle.$$

The expression in Eq. (1) includes a sum over final and average over initial magnetic and spin state degeneracy. The T matrix in Eq.(1) is the reaction amplitude, it couples the initial states ψ_i and the final states. T includes interaction between the incident and target electrons and the nucleus. It is the part of the TDCS that is the subject of approximation. The T-matrix element which represent the ionization amplitude is conveniently written in terms of distorted waves as

$$\left\langle k_1 k_2 \left| T \right| \psi_{nl} k_o \right\rangle = \left\langle X_1^{(-)}(k_1) X_2^{(-)}(k_2) \right| v_1 + v_3 - U \left| \psi_i X_0^{(+)}(k_0) \right\rangle \tag{2}$$

The electron-electron potential v_3 is responsible for ionization and U is the distorting potential. Use of orthogonality between orbital ψ_i and distorted wave for ejected electron eliminates v_1 and U from the expression of T matrix element. As mentioned earlier the initial state ψ_i contains an electron bound to the atom core with separation energy E_b. The distorted waves for the incident and scattered electrons, $X_0^{(+)}(k_0)$ and $X_1^{(-)}(k_1)$, respectively, are calculated in a potential $U = \left\langle \psi_i \middle| v_1 + v_3 \middle| \psi_i \right\rangle$.

Equation (1) in terms of direct and exchange amplitudes may be written as

$$\frac{d^3\sigma}{d\Omega_1\, d\Omega_2\, dE_1} = (2\pi)^4 \frac{k_1 k_2}{k_0} \sum_{av} \left(|f|^2 + |g|^2 - \mathrm{Re}(f^* g) \right) \tag{3}$$

where

$$f = \left\langle X_1^{(-)}(k_1, r_1) X_2^{(-)}(k_2, r_2) \middle| v_3 \middle| X_0^{(+)}(k_0, r_1) \psi_{nl}(r_2) \right\rangle, \tag{4}$$

$$g = \left\langle X_1^{(-)}(k_1, r_2) X_2^{(-)}(k_2, r_1) \middle| v_3 \middle| X_0^{(+)}(k_0, r_1) \psi_{nl}(r_2) \right\rangle \tag{5}$$

here $v_3 = \dfrac{1}{|r_1 - r_2|}$ is the interaction potential between the incident and target electrons responsible for the ionization, $X_0^{(+)}$ is the distorted wavefunction for the incident electron, $X_1^{(-)}$ and $X_2^{(-)}$ represent the distorted wavefunctions for the two outgoing electrons and each is orthogonalized with respect to

ψ_{nl}. Equations (4) and (5) are direct and exchange amplitudes for ionization from the (n, l) shell of the target atom where ψ_{nl} is the corresponding target orbital from which ionization is taking place and n and l are the principal and orbital quantum number respectively. We have used Hartree-Fock orbitals of Clementi and Roetti [24] for ψ_{nl}. The distorted wavefunction for the incident electron is generated in the equivalent local ground state potential of the atom whereas the distorted wavefunctions for the outgoing electrons are generated in the equivalent local ground state potential of the ion. For the work reported here we have made a careful check to ensure that the cross sections converge satisfactorily. The spin-averaged static-exchange potential [25] with the exchange potential taken in the equivalent local approximation is given as

$$V_E(r) = 0.5[E_0 + V_D(r) - \{[E_0 + V_D(r)]^2 - 2\pi\rho(r)^2\}^{1/2}] \tag{6}$$

where $\rho(r)$ is the electron density. The direct distorting potential $V_D(r)$ for the incident electron is obtained from the target radial orbital $u_{nl}(r)$ [26] as

$$V_D(r) = -\frac{Z}{r} + \sum_{nl} N_{nl} \int dr'[u_{nl}(r')]^2 / r_>, \tag{7}$$

where $r_>$ is the greater of r and r'. The details about the integration of the radial matrix element are described by McCarthy [26] and we have not given here for brevity. The equivalent local ground state potential V_{00}, which is the distorting potential, is the sum of exchange and direct potentials and is expressed as follows;

$$U = V_D(r) + V_E(r), \tag{8}$$

$$U = V_D(r) + 0.5[E_0 + V_D(r) - \{[E_0 + V_D(r)]^2 - 2\pi\rho(r)^2\}^{1/2}]. \tag{9}$$

We have modified the distorting potential (eq. (9)) used to calculate distorted waves $X_0^{(+)}, X_1^{(-)}$ and $X_2^{(-)}$ by addition of correlation-polarization potential V_{CP} to see the effect of electron correlation and target polarization on the trend of TDCS (DWBACPE calculations). The fundamental form of the short range correlation plus long range polarization potential may be approximated by means of local density functional theory (Padial and Norcross [27], Perdew and Zunger [28] and Yuan and Zhang [29]) as follows

$$\begin{aligned} V_{CP}(r) &= V_{SR}^{Corr}(r), \quad r \le r_0 \\ &= -\frac{\alpha_d}{2r^4}, \quad r > r_0 \end{aligned} \tag{10}$$

where α_d is dipole polarizability of the target and we have used short range correlation potential similar to [27].

3. Results and discussion

We present the results of triple differential cross section (TDCS) for electron impact ionization of Ne (2p) target in coplanar asymmetric geometry at incident electron energy 150 eV in Figure 1. We present the results of our standard DWBA and DWBACPE calculations by dash-dot-dotted and solid curves respectively. The other results plotted are DW2-RM (dashed curve) and DWBA-G (dotted curve) calculations [15]. The experimental results [15] (full circles) are shown for comparison. For a better comparison the DWBA-G results have been multiplied by a factor of 0.78 while plotting for Ne (2p) results. The experimental results have been normalized to the second binary lobe of our DWBACPE calculations for better comparison of TDCS shape since the absolute experimental results are not available. At scattering angle $\theta_1 = -15^0$ a deep minimum in the binary peak region is observed (Fig. 1a) in the experimental data and all the theoretical calculations including present

DWBA calculations. This corresponds to signature of the p orbital momentum distribution in the Bethe ridge kinematics.

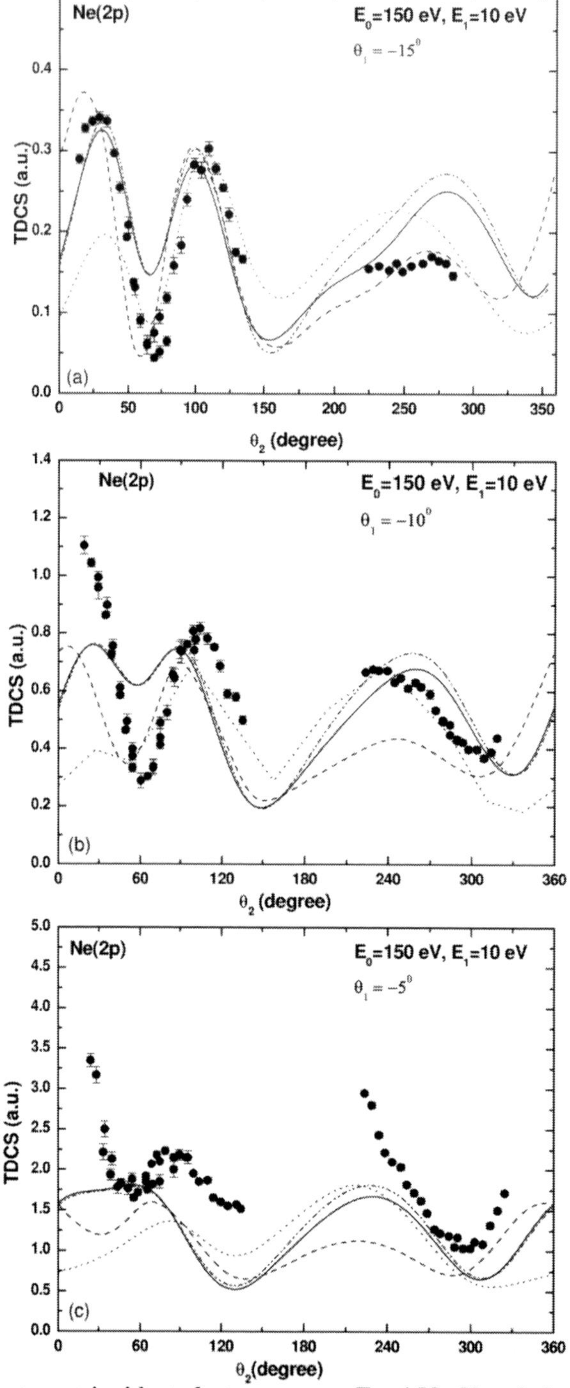

Figure 1: TDCS of Ne (2p) atom at incident electron energy E_0= 150 eV and ejected electron energy E_1 = 10 eV at various scattering angles θ_1 . Solid curve is our DWBACPE calculation, dash dot dotted curve is the standard DWBA calculation, dashed curve is DW2-RM and dotted curve is DWBA-G calculations [15]. The experimental results [15] are shown by full circles.

The depth of minimum decreases as scattering angle is decreased from $\theta_1 = -15^0$ to -10^0 and further to -5^0. Our DWBA and DWBACPE calculations are able to produce good agreement with the direction of primary and secondary binary peaks but there is a shift in the direction of recoil peak. The DWBACPE calculations which include correlation-polarization potential are able to improve the agreement with the experimental data. The DW2-RM [15] calculations are able to produce better binary to recoil peak ratio but the direction of binary peaks are not as observed in the experimental data. The DWBA-G [15] calculations overestimate the recoil peak whereas underestimate the binary peak at the scattering angles $\theta_1 = -15^0$ and -10^0. Our DWBACPE results provide a good agreement in terms of direction of binary peak but there are still certain discrepancies in terms of direction of recoil peak and the binary to recoil peak ratio for the ionization of Ne (2p) target.

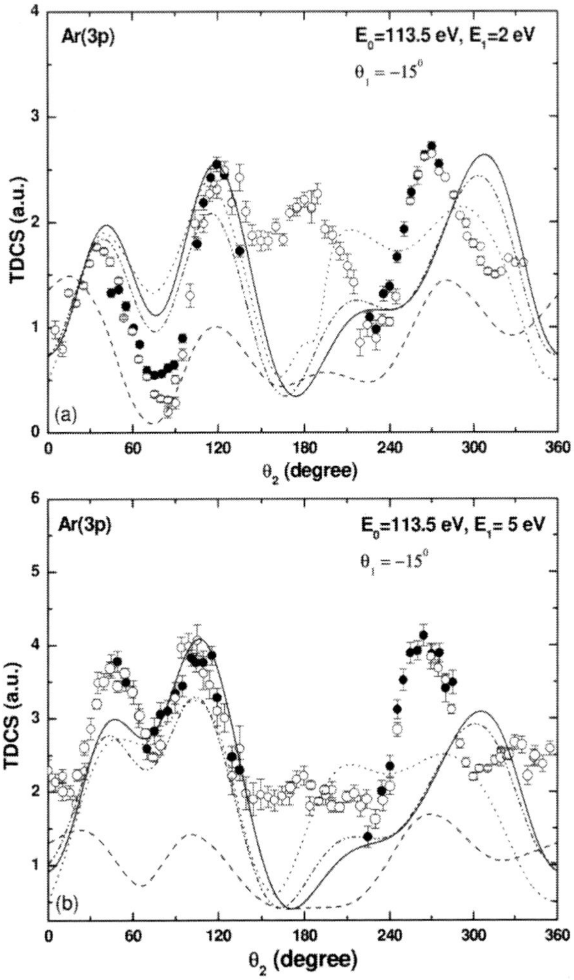

Figure 2: TDCS of Ar (3p) atom at incident electron energy E_0= 113.5 eV and ejected electron energy E_1 = (a) 2 eV and (b) 5 eV at various scattering angles θ_1. Solid curve is our DWBACPE calculation, dash dot dotted curve is standard DWBA calculation, dashed curve is DW2-2st.-MC and dotted curve is 3DW calculations taken from [17]. The solid circles are experimental results of [10] and the open shapes are the results of [17].

The results of TDCS in the coplanar asymmetric (e, 2e) processes on Ar (3p) atoms are presented in Figure 2. The DWBACPE and standard DWBA calculations are depicted by the solid and dash-dot-dotted curves respectively. The other calculations plotted are DW2-2state-MC (dashed curve) and 3DW (dotted curve) taken from [17]. The solid circles are experimental results of [10] and the open shapes are experimental results of [17]. The experimental data have been normalized to our DWBACPE calculations at the secondary binary peak. We observe that at ejected electron energy 2 eV (Fig. 2a) the DWBACPE, standard DWBA and 3DW calculations are able to provide good agreement with both the binary peaks, however all of these calculations underestimate the dip near the momentum transfer direction. As it has been mentioned in the discussion of Ne (2p) ionization, the dip near the momentum direction is signature of ionization from p-orbital. It is interesting to observe that the inclusion of correlation-polarization potential improves the agreement with the measurements in the binary peak region. The agreement in the recoil peak direction is not so good in the present DWBACPE and standard DWBA calculations. We believe that that inclusion of PCI in the DWBA calculations may improve the agreement with the experimental results. The DW2-2state-MC calculation gives good agreement with the experimental results in the second binary peak region however it overestimates the first binary peak and the direction is also not correct. The DW2-2state-MC calculation gives better agreement in the recoil peak region in comparison to the other calculations at ejected electron energy 2 eV.

In case of ionization from the 3p shell of argon at ejected electron energy 5 eV (Fig. 2b) the present DWBACPE, standard DWBA and 3DW calculations are able to give nearly correct trend of TDCS in the binary peak region, however there are significant discrepancies in terms of magnitude and direction in the recoil peak region. The DW2-2state-MC calculation is able to produce most of the trends of TDCS at ejected electron energy 5 eV.

The results of triple differential cross sections (TDCS) for the electron impact ionization of Xe (5p) are presented in Figure 3. The results of our standard DWBA and DWBACPE calculations are depicted by dash-dot-dotted and solid curve respectively. The other results shown are DW2-RM (dashed curve) and DWBA-G (dotted curve) calculations [15]. The experimental results [15] (full circles) are also shown for comparison. The DWBA-G results have been multiplied by a factor of 1.35 for a better comparison. The experimental results have been normalized to the binary peak of our DWBACPE calculations for better comparison of TDCS shape. As in the case of ionization of Ne (2p) target, the ionization of Xe (5p) at $\theta_1 = -15^0$ corresponds to the Bethe ridge condition. The trace of p-orbital momentum distribution is very weak for the Xe target at $\theta_1 = -15^0$ and it disappears as scattering angle is reduced (Figs. 3b, 3c). Our DWBACPE calculations which include correlation-polarization potential give an improved agreement with the experimental data in the binary peak direction in comparison to standard DWBA and other available theoretical calculations. However, our calculations and all other available theoretical calculations underestimate the TDCS in the recoil peak region.

We have presented the results of TDCS of the (e, 2e) processes on Ne (2p), Ar (3p) and Xe (5p) targets in the modified distorted wave Born approximation formalism by including the correlation-polarization potential (as a function of electron density) in the standard DWBA formalism. We have also compared the results of our present calculations with the recent available theoretical and experimental results. We observe that there is mixed degree of agreement between various theoretical and experimental results. Particularly, our DWBACPE calculation is able to produce good agreement in the binary peak region for the ionization from Ar (3p) and Xe (5p) targets, however the agreement in case of ionization from Ne (2p) target is not so good. This may be due to increase in the polarizabilities from Ne to Ar and Xe targets so the effect of polarization potential on the trend of TDCS is more significant for Ar and Xe targets. We conclude that polarization of target is an important effect at the incident electron energies considered in the present investigation. Still there are certain discrepancies in the trend of TDCS of noble gas targets Ne, Ar and Xe which require further investigation of (e, 2e) processes on these targets. We strongly believe that the inclusion of PCI, second order processes and better treatment of polarization phenomenon may improve the trends of TDCS.

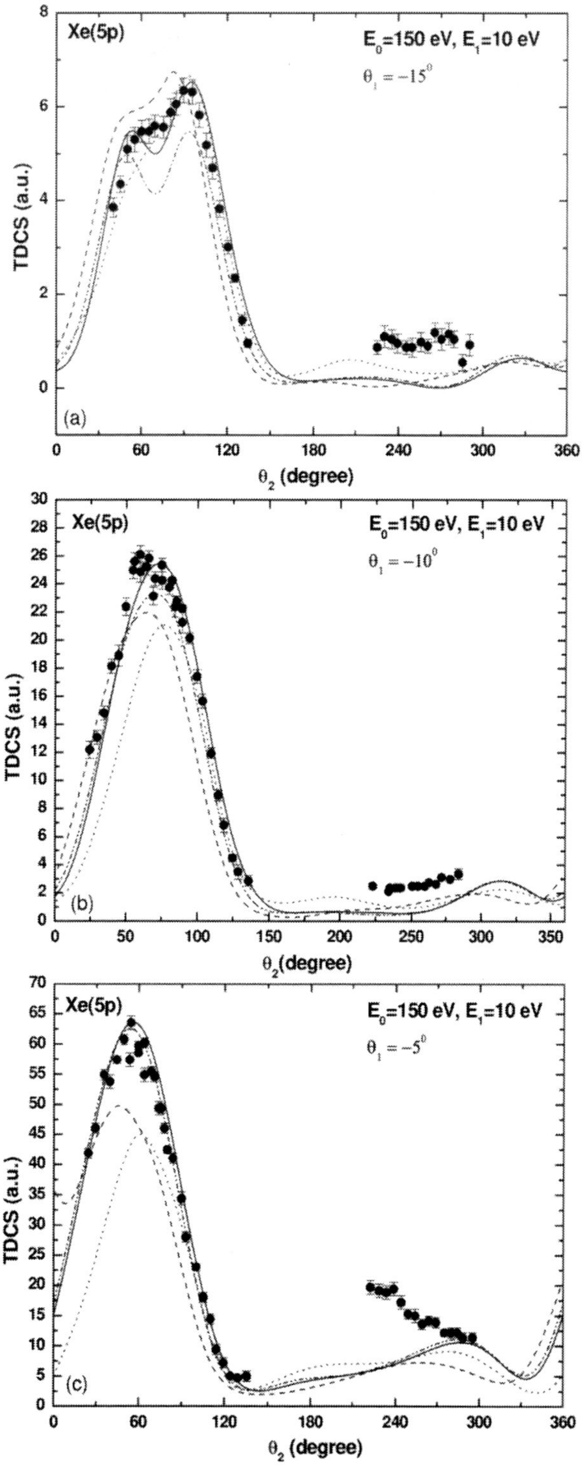

Figure 3: TDCS of Xe (5p) atom at incident electron energy E_0= 150 eV and ejected electron energy E_1= 10 eV at various scattering angles θ_1. Solid curve is our DWBACPE calculation, dash dot dotted

curve is the standard DWBA calculation, dashed curve is DW2-RM and dotted curve is DWBA-G calculations [15]. The experimental results [15] are shown by full circles.

References

[1] Ehrhardt H, Schulz M, Tekaat T and Willmann K 1969 Phys. Rev. Lett. **22**, 89.
[2] Rescigno T N, Baertschy M, Isaacs W A and McCurdy C W 1999 Science **286,** 2474.
[3] Bray I, Fursa D V, Kheifets A S and Stelbovics A T 2002 J. Phys. B **35**, R117.
[4] Whelan C T, Allan R J, Walters H R J and Zhang X 1993 (e, 2e) and related processes ed. C. T. Whelan et al. (Dordrechat: Kluwer) p1.
[5] McCarthy I E and Weigold E 1976 Phys. Rep. **C27**, 275.
[6] Byron Jr F W and Joachain C J 1989 Phys. Rep. **179**, 211.
[7] Nakel W and Whelan C T 1999 Phys. Rep. **315**, 409.
[8] Waterhouse D K, McCarthy I E and Williams J F 1998 Phys. Rev. A **57**, 3565.
[9] Haynes M A and Lohmann B 2000 J. Phys. B **33**, 4711.
[10] Haynes M A and Lohmann B 2001a Phys. Rev. A **64**, 044701.
[11] Haynes M A and Lohmann B 2001b J. Phys. B **34,** L131.
[12] Biava D A, Saha H P, Engel E, Dreizler R M, McEachran R P, Haynes M A, Lohmann B, Whelan C T and Madison D H 2002 J. Phys. B 35, 293.
[13] Haynes M A, Lohmann B, Prideaux A and Madison D H 2003 J. Phys. B 36, 811.
[14] Stevenson M, Leighton G J, Crowe A, Bartschat K, Vorov O K and Madison D H 2005 J. Phys. B **38**, 433.
[15] Stevenson M, Hargreaves L R, Lohmann B, Fursa D V, Bartschat K and Kheifets A 2009 Phys. Rev. A **79**, 012709.
[16] Stevenson M and Lohmann B 2006 Phys. Rev. A **73** 020701.
[17] Stevenson M and Lohmann B 2008 Phys. Rev. A **77** 032708.
[18] Purohit G, Vinod Patidar and Sud K K 2009, J. Elect. Spec. and Rel. Phenom. **175**, 1.
[19] Prideaux A and Madison D H 2003 Phys. Rev. A **67**, 052710.
[20] Bartschat K and Vorov O 2005 Phys. Rev. A **72**, 022728.
[21] Kheifets A S, Naja A, Staicu Casagrande E M and Lahmam-Bennani A 2008 J. Phys. B **41**, 145201.
[22] Prideaux A, Madison D H and Bartschat K 2005 Phys. Rev. A **72,** 032702.
[23] Otranto S 2009 Phys. Rev. A **79** 012705.
[24] Clementi E and Roetti C 1974 At. Dat. Nucl. Dat. Tab. 14.
[25] Furness J B and McCarthy I E 1973 J. Phys. B **6**, 2280.
[26] McCarthy I E 1995 Aust. J. Phys. **48** 1.
[27] Padial N T and Norcross D W 1984 Phys. Rev. A **29**, 1742.
[28] Perdew J P and Zunger A 1981 Phy. Rev. B **23**, 5048.
[29] Yuan J and Zhang Z 1989 J. Phys. B **22**, 2751.

Two-dimensional Penning Ionization Electron Spectroscopy of CH_3I and CH_2I_2 by $He^*(2^3S)$ Metastable Atoms

Naoki Kishimoto, Andriy Borodin, Takuya Horio, Masakazu Yamazaki, and Koichi Ohno

Department of Chemistry, Graduate School of Science, Tohoku University, Aramaki, Aoba-ku, Sendai 980-8578, Japan

E-mail: n-kishi@m.tains.tohoku.ac.jp

Abstract. Penning ionization of iodomethane (CH_3I) and diiodomethane (CH_2I_2) upon collision with metastable $He^*(2^3S)$ atoms was studied by collision-energy/electron-energy resolved two-dimensional Penning ionization electron spectroscopy (2D-PIES) over a wide range of collision energies (20 – 350 meV). Collision energy dependence of partial ionization cross sections (CEDPICS) indicates that the interaction with $He^*(2^3S)$ atoms around the valence molecular orbitals is attractive at low collision energies except for the only σ_{CH} molecular orbital region of iodomethane which has a repulsive character in the normal collision energy range (70 – 350 meV). Large bands at lower electron energies (3.7 – 5.5 eV) in PIES spectra were ascribed to autoionization from a transient ion-pair. The related entrance ionization channel has an attractive character. The excitation transfer, which is responsible for the I** autoionization, takes place in regions close to carbon atoms and is repulsive towards $He^*(2^3S)$ in the normal collision energy range. The superexcited complex of $(He–CH_2I_2)^{**}$ autoionizes via the excitation transfer occurring close to the σ_{CI} regions of the targets.

1. Introduction

Unraveling the dynamics of chemical reactions is of fundamental importance. In the case of chemi-ionization process known as the Penning ionization ($A^* + M \rightarrow A + M^+ + e^-$) [1], the ionization of a target atom or molecule (M) occurs by collision with a metastable atom (A*). Important information on the dynamics of this process can be obtained by resolving the collision energy between reactants and analyzing the energy distributions of ejected electrons. Usually, ionic states can be related to valence molecular orbitals (MOs). In the Penning ionization, an electron in MO having higher electron density outside the boundary surface of M is transferred to the inner-shell orbital of the probe atom A*, which results in a high-intensity band in Penning ionization electron spectroscopy (PIES) according to the electron exchange model [2].

Extensive studies have been made for Penning ionization with a metastable A* beam, and the collision energy dependence of total ionization cross sections (σ_T) by product-ion detection [3] as well as the scattering angular distribution [4] have been measured. Olson showed that the experimental collision energy dependence of the total ionization cross section for $He^*(2^{1,3}S)$ + Ar was well reproduced by semiempirical calculations [5]. In collision energy region lower than 100 meV, the ionization cross section decreases with the increase of collision energy, and then starts to increase to 15 $Å^2$ around a collision energy of 10 eV. The general features of the collision energy dependence of

Penning ionization cross sections can be classified into three types as follows. The first type (I) shows negative collision energy dependence, and it was related to the attractive interactions. If the long-range attractive part of the interaction potential $V^*(R)$ as a function of inter atomic distance R, with the potential well ε, plays a dominant role and its function form is the type of

$$V^*(R) \propto R^{-s}, \qquad (1)$$

$\sigma_T(E_c)$ can be expressed by

$$\sigma_T(E_c) \propto E_c^{-2/s}. \qquad (2)$$

This formula for Langevin-type cross section is derived from the impact parameter value at which orbiting occurs. The potential parameter s can be related to interaction forces such as dispersion forces or induced dipole interactions (s = 6) and ionic interactions (s = 4). In the second type (II), the repulsive part of interaction potential governs the collision energy dependence. In the simple theoretical model introduced by Illenberger and Niehaus [6], $\sigma_T(E_c)$ can be expressed as

$$\sigma_T(E_c) \propto [\ln (A/E_c)]^2 (E_c/A)^{(b/d)-(1/2)}, \qquad (3)$$

where simple analytical forms of the interaction potential $V^*(R)$ and the ionization width $\Gamma(R)$ are assumed by

$$V^*(\mathrm{R}) = A \exp(-dR), \qquad (4)$$

and

$$\Gamma(R) = B \exp(-bR). \qquad (5)$$

When the minor contribution of the first factor in Eq. (3) is neglected, the collision energy dependence can be related to d and b via the slope m in a plot of log σ_T-log E_c by

$$m = (b/d) - (1/2). \qquad (6)$$

Here, d and b are effective decay parameters of repulsive potential wall (V^*) and ionization width (Γ), respectively. The third type (III) corresponds to the hard core limit of Eq. (6); the parameter d becomes considerably large and slope m of a log σ_T-log E_c plot is nearly -1/2. In this case, the effective interaction time deceases with the increase of E_c.

Combining an electron energy analysis with a velocity selection of the metastable incoming atomic beams [7,8] enables us to measure the collision energy dependence of partial ionization cross-sections (CEDPICS) by collision-energy/electron-energy-resolved two-dimensional Penning ionization electron spectroscopy (2D-PIES) [8,9]. Since the asymptotic decay of all MOs is directly related to the lowest ionization energy (I_0) of the target molecule [10-12] by an expression of $\exp\{-2(2I_0)^{1/2}R\}$ for the distance R, the radial dependence of the ionization width in Eq. (5) for all MOs should be the same. However, the reaction probability to each ionic state was found to show a difference in the collision energy dependence with the change of E_c [7]. This difference can be ascribed to the highly anisotropic interactions around the target molecule, where trajectories of the metastable atoms are deflected variously around the reactive region depending on the collision energy, the impact direction, and the impact parameter. For the target molecule, since the electron distributions of individual MOs are more or less localized on the spatial parts of the molecule, different CEDPICS for various ionic states can be connected to the anisotropy of interaction between a target molecule M and a metastable atom A* [7-9]. In the case of positive CEDPICS, the metastable atom with larger kinetic energy reaches the inner region of the target against a repulsive interaction. Thus, a positive CEDPICS implies that the interaction around the region where a corresponding MO spatially extends is repulsive. On the other hand, negative CEDPICS can be related to an attractive interaction around the region where the corresponding MO distributes, since the metastable atom is deflected towards the target molecule by the attractive interaction. In this case, the extent of the deflections decreases with the increase of collision energy.

The interaction between a metastable atom and a target molecule does not necessarily lead to a direct Penning process, and two further sequences also can lead to electron ejection when a molecular target collides with a metastable $He^*(2^3S)$ atom; (i) Electron ejection can be achieved through autoionization via an ionic reaction channel. In this channel, the He 2s electron transfers to the target which results into the formation of an ionic pair He^+–M^-. Then the M^- component autoionizes in a subsequent step, and the ionic pair He^+–M^- transforms into He, M^+, and e^-. Dissociation of M into some fragments can occur. (ii) A more complicated autoionization reaction channel can involve an electronic excitation transfer: the He 2s electron transfers to an unoccupied target molecular orbital accompanied by a filling of the He 1s hole orbital by an electron of M. The produced intermediates are electronically excited M* species which can autoionize; or alternatively, undergo dissociation in excited fragments that can also autoionize. Moreover for some substances such as compounds containing halogen atoms, complicated and various autoionization processes can also play an essential role and lead to spectral bands with large ionization cross section. Broad intensive bands were observed in the PIES spectra of Cl_2, Br_2, and I_2 [13,14]. It is interesting that for heteronuclear halogen compounds (ICl and IBr), the autoionization structure can be in general represented as a simple superposition of corresponding bands of related homonuclear compounds (Cl_2, Br_2, and I_2) [14]. For HCl, HBr, and HI, the excitation transfer has been observed [15-17], and PIES measurements with high electron energy resolution led to observation of a series of small peaks in the low electron energy region (0 – 3.7 eV), which was assigned to the autoionization of the Rydberg series of the dissociated halogen atom.

Our group has applied 2D-PIES to some methyl compounds containing halogen such as bromo methanes, CH_2ClI, CH_2BrI, $CHBrCl_2$ in order to investigate the reaction dynamics and anisotropic interactions with He* atoms [18-20]. In this study, we have investigated the anisotropic interaction with He* atoms and autoionization process for Penning ionization of CH_3I and CH_2I_2 with the 2D-PIES technique.

2. Methods

The experimental apparatus used for the measurements is described in detail in previous papers [8,9]. Briefly, metastable He*(2^3S) atoms were produced by a discharge nozzle source. In addition, a water-cooled helium discharge lamp was used for quenching the metastable He*(2^1S) atoms generated as a byproduct in the source. To obtain the collision-energy-resolved spectra, a chopper disk rotating with a frequency of about 350 Hz modulated the beam of the metastable helium atoms. For the UPS measurements, the apparatus was equipped with an additional source of photons from the He I resonance line (584 Å, 21.22 eV). The interaction between the molecules under investigation and the metastable He*(2^3S) atoms as well as with the He I photons occurred in a small collision cell (20 mm in diameter). A hemispherical electrostatic deflection type analyzer connected to the collision cell was used to measure the kinetic energy of the ejected electrons. The transmission curve for the energy analyzer was calibrated on the basis of the relative peak intensities for various compounds with respect to known UPS data [21,22].

The collision energy dependence of the partial ionization cross-section was obtained for a specific ionic state in 2D-PIES accumulated as functions of two parameters; the electron kinetic energy (E_e) and the collision energy (E_c) between A* and M. The 2D-PIES measurements in the low collision energy range have been performed with slow metastable He* atoms produced by discharge between the skimmer and a nozzle cooled by liquid nitrogen [23]. The retarding type electron analyzer based on the 'magnet bottle effect' has been utilized for collecting electronic counts at lower collision energy region. The velocity distribution of the metastable He* beam was monitored by secondary electrons emitted from a stainless steel plate inserted in the reaction cell. Then we obtained CEDPICS in lower (20-100 meV) and higher (60-300 meV) collision energy range.

We performed ab initio molecular orbital calculations with the Gaussian 03 package of programs [24]. The geometries of the neutral target molecules were obtained by microwave spectroscopy for CH_3I [25] and optimization with second-order Møller-Plesset perturbation theory (MP2) with 6-311G** basis functions. In the electron density contour maps, thick solid curves indicate the repulsive molecular surface approximated by atomic spheres of van der Waals radii [26]. For the ionization potential energies, the outer valence Green's function (OVGF) [27] calculations were performed. Based on the well-known resemblance between He*(2^3S) and Li(2^2S) [3], a Li(2^2S) atom in the ground state was utilized in place of a He*(2^3S) atom for model potential surface calculations for the interacting system He*-M. The basis set superposition error was corrected by counterpoise method [28].

According to the electron exchange model [2,8] of Penning ionization process, the probability of partial electron emission intensities for ionization from selected electronic states are thought to be proportional to the overlap between the unoccupied 1s MO of the He*(2^3S) atomic probe with the ionized target MO (φ_i) within the subspace (Ω) outside the collision boundary surface:

$$(EED)_i = \int_\Omega |\varphi_i(r)|^2 dr \qquad (1)$$

In practice, the collision boundary surface is equated to a rigid van der Waals surface assuming a hard sphere depiction [2,8]. Exterior electron densities values (EED), from which partial ionization cross sections can be estimated, have been calculated at the HF/6-311G** level. In that way, on the basis of known (from UPS or ab initio calculations) ionization potentials we can simulate the Penning ionization electron spectra from investigated target. By comparison of this spectrum with real PIES data we can distinguish bands caused by other ionization channels.

3. Results and Discussion

3.1. Valence ionic states observed by UPS and PIES

It is well-known that the first and second bands in the UPS of iodomethane (CH_3I) show spin-orbit splitting ($5P_{1/2}$ and $5P_{3/2}$) of degenerate 9e orbitals [30], which emerges as two peaks in almost equal intensities at 9.5 and 10.8 eV in the UPS spectrum (Figure 1(a)). The strong intensity of bands 1 and 2 in PIES indicates spatial extending of large electron density outside of the collision limit (molecular surface) as shown in the EED spectrum. Band 3 was assigned to $12a_1$ MO which also has relation with the iodine lone-pair electron. Degenerate 8e orbitals having σ_{CH} character were observed as bands 4, 5 for which broadening was induced by the Jahn-Teller effect. Observed band intensity of band 3 and bands 4, 5 is not so large as bands 1, 2, which can be ascribed to bonding character of $12a_1$ MO between the iodine atom and the methyl group (σ_{CI}) and 8e MO (σ_{CH}) with small spatial extending outside the molecular surface.

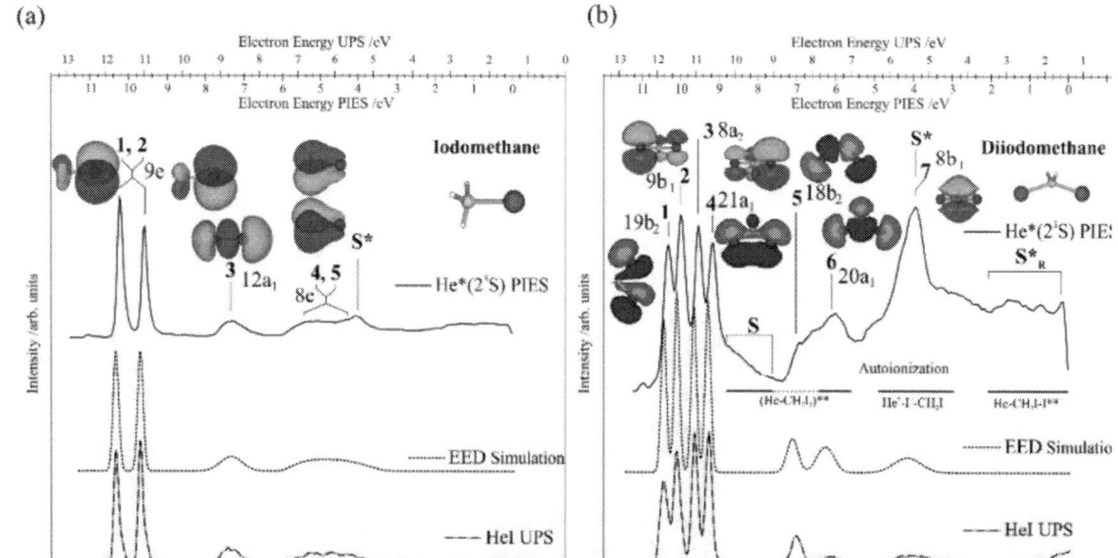

Figure 1. He I UPS, He*(2^3S) PIES, and EED simulation of (a) CH_3I and (b) CH_2I_2.

The four bands corresponding to lone-pair MOs of diiodomethane (CH_2I_2) were assigned to $19b_2$, $9b_1$, $8a_2$, and $21a_1$ in the order calculated with the OVGF method (Figure 1(b)). Our assignment is consistent with an earlier UPS study [30]. However, ionization energy values of these lone-pair MOs of CH_2I_2 can be difficult to estimate by the quantum chemical theory because of spin-orbit splitting effects in ionization from iodine lone-pair orbitals. Next three bands (bands 5, 6, and 7) were observed in He I UPS spectrum at the electron energy of 4–9 eV, and they were assigned to σ_{CI} and σ_{CH} MOs based on the results by the OVGF calculation as shown in Table 1. Contrary to the UPS and the EED spectrum (Figure 1(b)), these bands show strong intensity in PIES. The reason for this strong intensity in PIES will be discussed in Section 3.3.

3.2. Collision-energy-resolved measurements of PIES

Negative slope of CEDPICS for CH_3I (Figure 2 (a)) for bands 1 and 2 (n_I, m(higher) = −0.60, m(lower) = −0.40) can be ascribed to attractive interaction between the target molecule and He*(2^3S) atoms around the n_I orbital region. It is known that slower He* atoms can penetrate into the inner reactive region than the faster He* atoms because of the lower centrifugal barrier, which can lead to negative CEDPICS [7,8]as described in Section 1. Attractive interaction was found by the model potential calculation for the perpendicular direction around the C-I axis (Figure 3(a)). The small well depth (ca. 10 meV) of the interaction potential may be due to the Li-model potential calculation with insufficient basis set for the I atom. On the other hand, repulsive interaction was obtained around the σ_{CH} orbital region, which is consistent with the small positive slope of CEDPICS at higher collision energy region (60-300 meV) for bands 4,5 (σ_{CH}, m = +0.04). The bending shape of CEDPICS for bands 4,5 may be influenced by the attractive interaction around the iodine atoms in spite of the ionization from σ_{CH} MOs for slower metastable He^* atoms. The negative slope of CEDPICS for band 3 (σ_{CI}, m(higher) =

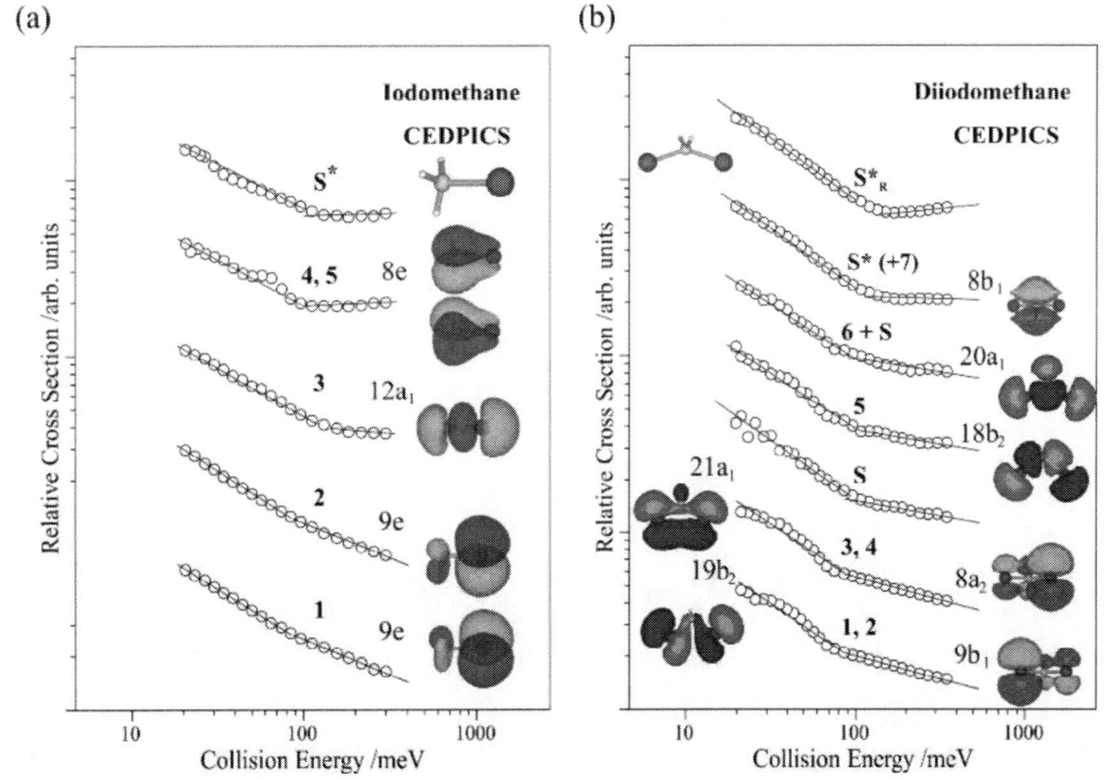

Figure 2. CEDPICS of (a) CH_3I and (b) CH_2I_2 by collision with He*(2^3S) atoms.

–0.09) is smaller than that for n_I bands, which is consistent with the weaker interaction potential along the C-I axis rather than the perpendicular direction around the C-I bond. For this band, weak bending shape of CEDPICS was observed.

For the case of CH_2I_2 (Figure 2(b)), similar negative collision energy dependence was observed for bands 1,2 and bands 3,4. Influence of attractive interactions was calculated around the iodine atoms of CH_2I_2 (Figure 3(b)) rather than CH_2 group. It is interesting that the shape of CEDPICS was bending at lower collision energy region. The CEDPICS for other bands will be discussed in the following Section 3.3.

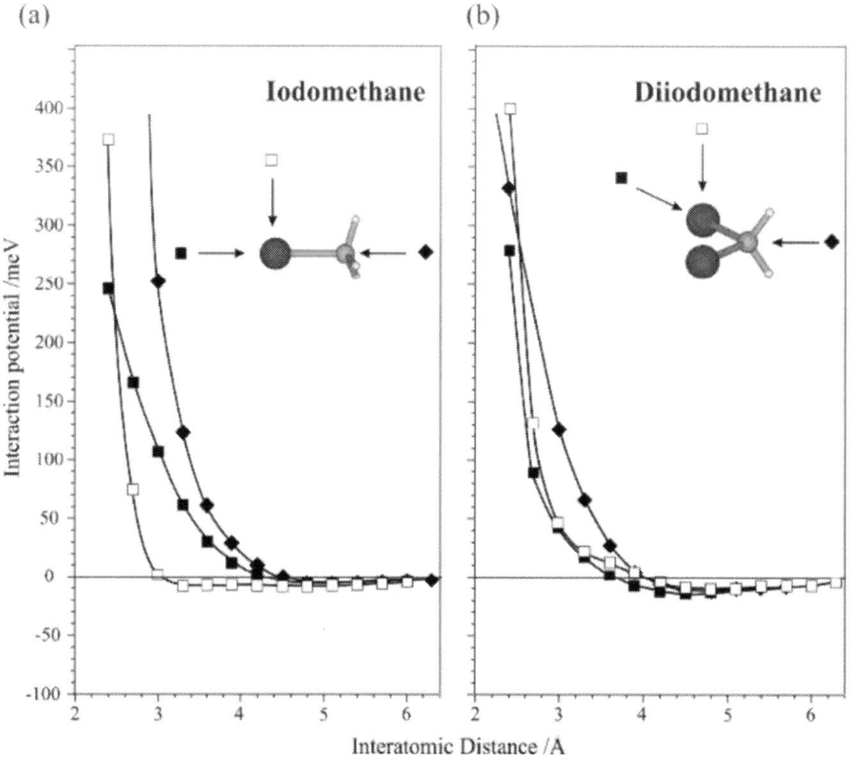

Figure 3. Model potential curves for M-Li of (a) CH_3I and (b) CH_2I_2 as a function of interatomic distance between Li and I or C atoms.

3.3. Autoionization bands

Former studies on iodine compounds have pointed out the presence of autoionization bands (S^* in this study) in PIES of HI [17], I_2, IBr, and ICl [14]. The position of S^* bands depend on the energy curve for the dissociation of the intermediate ionic pair $He^+\cdots I^-$-R [17]. For CH_3I, small S^* band can be found around E_e = 4 eV in PIES, and no counter part band structure was observed in He I UPS. The strong S^* band of CH_2I_2 (Figure 1(b)) relating to the formation of intermediate ionic pair ($He^+\cdots I^-$-CH_2I) is located at electron energy E_e = 3.94 eV in PIES, which is close to the diatomic iodine molecule cases (3.75 eV for HI [17], 3.2 eV for I_2, IBr, and ICl [14]). Band shoulder with no strong counter part was observed at E_e = 5.1 eV in He I UPS for CH_2I_2 in this study and other diatomic iodine molecules (I_2, IBr, and ICl) [14]. Observed intensity of S^* band is extensively large compared with band 7 in the EED spectrum (Figure 1(b)). The shape of CEDPICS for the S^* band of CH_3I and CH_2I_2 is bending with large negative slope at lower collision energy region and small positive slope at higher

Table 1. Ionization potentials (eV) by He I UPS and theoretical calculations, band assignments, EED values (%) and slope parameter (*m*) of CEDPICS at lower (20-100 meV) and higher (60-300 meV) collision energy region for CH_3I and CH_2I_2

Band	IP	Koopmans	OVGF	EED	*m*(lower)	*m*(higher)
CH_3I						
1	9.48	9.85(9e, n_I)	9.50 (0.93)[a]	6.51	−0.60	−0.40
2	10.11	9.85(9e, n_I)		6.51	−0.62	−0.40
3	12.40	12.67(12a_1, σ_{CI})	12.31 (0.92)	3.00	−0.54	−0.09
4	15.1	16.59(8e, σ_{CH})	15.28 (0.89)	3.26	−0.53	+0.04
5		16.59(8e, σ_{CH})		3.26		
S^*	(15.8)				−0.50	+0.03
CH_2I_2						
1	9.36	9.86(19b_2, n_I)	9.36(0.92)	5.45	−0.65	−0.26
2	9.70	10.24(9b_1, n_I)	9.68(0.92)	6.52	−0.65	−0.26
3	10.14	10.39(8a_2, n_I)	9.82(0.92)	6.02	−0.65	−0.26
4	10.50	10.52(21a_1, σ_{CI})	9.95(0.92)	6.23	−0.65	−0.26
S	(11.2-12.6)				−0.68	−0.17
5	12.68	13.23(18b_2, σ_{CI})	12.54(0.91)	2.48	−0.62	−0.16
6	13.48	14.43(20a_1, σ_{CI})	13.39(0.89)	2.82	−0.62	−0.15
7	15.50	17.65(8b_1, σ_{CH})	15.82(0.88)	2.62	−0.68	+0.02
S^*	(15.9)				−0.68	+0.02
S^*_R	(18.1-19.8)				−0.70	+0.09

[a] Pole strengths are shown in the parentheses.

collision energy region, which can be explained by the collision dynamics around the crossing point of interaction potential energy curves for intermediate ionic pair $He^+\cdots I^-$-R and normal Penning ionization collision of $He^* + M$. These results indicate that metastable atoms at lower collision energy are strongly influenced by the attractive force on the interaction potential energy curve for the ionic pair. In addition, small positive collision energy dependence was observed for the normal Penning ionization process around the C-I axis at higher collision energy region than $E_c \sim 150$ meV.

Autoionization bands like band S for CH_2I_2 in this study were observed at the shoulder of valence bands for $CHBr_3$ [20] and HX [14-17], and they were assigned to ionization from superexcited states of a complex $(He–M)^{**}$ to highly excited vibrational states of ion. The large negative CEDPICS for band S of CH_2I_2 can be ascribed to attractive interaction around the iodine at generating the superexcited states. The observed intensity of band 6 for CH_2I_2 is also different from the EED spectrum, and observed CEDPICS for band 6 is very similar to that for band 5. Judging from the similar MO distribution of 18b_2 and 20a_1 and CEDPICS for bands 5 and 6, the additional signals around the band 6 also can be assigned to autoionization (S band) related to bands 5 and 6.

For lower electron energy region in PIES of HI, sharp lines by autoionization of $[^1S_0]$ns and $[^1S_0]$np series Rydberg I^{**} atom into the $^1D_2$ I^+ state were observed [17]. Thus, signals in lower electron energy region than 3.6 eV in PIES of CH_2I_2 may be ascribed to Rydberg I^{**} (S^*_R) atoms induced by the dissociation of the C-I bond of $(CH_2I\text{-}I)^{**}$. At the low electron energy region in PIES for CH_3I, electronic counts are low, and the band structure was not significant. The small or indistinct S^*_R band for CH_3I may be related to small transition probability to excited $(CH_3\text{-}I)^{**}$ states by the approach of $He^*(2^3S)$ atoms. The positive slope of CEDPICS for the S_R^* band of CH_2I_2 at higher collision energy region ($m = +0.09$) may indicate collision dynamics for generating the $(CH_2I\text{-}I)^{**}$ states. When we assume the anisotropy of interaction potential for the He^*-CH_2I_2 system in the excitation transfer from He^* to CH_2I_2, the He^* approaching direction can be dominant around the σ_{CH} bond region.

Acknowledgements
This research was supported by a Grant-in-Aid for Scientific Research from the Japan Society for Promotion of Science (JSPS). The authors (B. A., H. T. and M. Y.) were supported by the Postdoctoral Fellowship for Foreign Researchers (B. A.) and Young Scientists (H. T. and M. Y.) by JSPS.

References

[1] Penning F M, 1927 Naturwissenschaften **15** 818.
[2] Ohno K, Mutoh H, and Harada Y 1983 J. Am. Chem. Soc. **105** 4555.
[3] Niehaus, A 1981 Adv. Chem. Phys. **45**, 399, and references cited therein.
[4] Siska P E, 1993 Rev. Mod. Phys. **65**, 337, and references cited therein.
[5] Olson R E, 1972 Phys. Rev. A **6**, 1025.
[6] Illenberger E, Niehaus A, 1975 Z. Phys. B **20**, 3590.
[7] Ohno K, Takami T, Mitsuke K, Ishida T, 1991 J. Chem. Phys. **94** 2675.
[8] Ohno K, 2004 Bull. Chem. Soc. Jpn. **77** 887.
[9] Kishimoto N, Ohno K, 2007 Int. Rev. Phys. Chem. **26** 93.
[10] Handy N C, Marron M T, Silverstone H J, 1969 Phys. Rev. **180**, 45.
[11] Morrell M M, Parr R G, Levy M, 1975 J. Chem. Phys. **62**, 549.
[12] Katriel J, Davidson E R, 1980 Proc. Natl. Acad. Sci. USA **77**, 4403.
[13] Kilschat W, Morgner H, 1983 Z. Phys. A. **312** 305.
[14] Beckmann K, Leisin O, Morgner H, 1986 Mol. Phys. **59** 829.
[15] Yencha A, Ganz J, Ruf M -W, Hotop H, 1989 Z. Phys. D: At., Mol. Clusters **14** 57.
[16] Yencha A, Ruf M -W, Hotop H, 1991 Z. Phys. D: At., Mol. Clusters **21** 113.
[17] Yencha A, Ruf M -W, Hotop H, 1994 Z. Phys. D: At., Mol. Clusters **29** 163.
[18] Tian S X, Kishimoto N, Ohno K, 2003 J. Phys. Chem. A **107** 485.
[19] Tian S X, Kishimoto N, Ohno K, 2003 J. Phys. Chem. A **107** 2137.
[20] Kishimoto N, Matsumura E, Ohno K, Deleuze M S, 2004 J. Chem. Phys. **121** 3074.
[21] Gardner J L, Samson J A R, 1976 J. Electron Spectrosc. Relat. Phenom. **8** 469.
[22] Kimura K, Katsumata S, Achiba Y, Yamazaki T, Iwata S, Handbook of HeI Photoelectron Spectra of Fundamental Organic Molecules (Japan Scientific: Tokyo, 1981).
[23] Horio T, Yamazaki M, Maeda S, Hatamoto T, Kishimoto N, and Ohno K 2005 J. Chem. Phys. **123** 194308.
[24] Frisch M. J., *et al.* Gaussian 03 (Gaussian, Inc., Pittsburgh, PA, 2003).
[25] Mallinson P D, 1975 J. Mol. Stectrosc. **55** 94.
[26] Pauling L, The Nature of the Chemical Bond (Cornell University, Ithaca, NY, 1960).
[27] Ortiz J V, 1996 J. Chem. Phys. **104** 7599.
[28] Boys S B, Bernardi F, 1970 Mol. Phys. **19** 553.
[29] Eland J H, Photoelectron Spectroscopy: An introduction to Ultraviolet Photoelectron Spectroscopy in Gas Phase (Butterworth, London, 1974).
[30] von Niessen W, Åsbrink L, Bieri G, 1982 J. Electron Spec. Relat. Phenom. **26** 173.

Radiative charge transfer and interatomic Coulombic decay following direct double photoionization of neon dimers

I Higuchi[1], T Ouchi[2], K Sakai[2], H Fukuzawa[2], X-J Liu[2], K Ueda[2], H Iwayama[3], K Nagaya[3], M Yao[3], D Ding[4], D Zhang[4], Y Tamenori[1] and N Saito[5]

[1] Japan Synchrotron Radiation Research Institute, Sayo, Hyogo 679-5198, Japan
[2] Institute of Multidisciplinary Research for Advanced Materials, Tohoku University, Sendai 980-8577, Japan
[3] Department of Physics, Graduate School of Science, Kyoto University, Kyoto 606-8502, Japan
[4] Institute of Atomic and Molecular Physics, Jilin University, Changchun 130012, P. R. China
[5] National Institute of Advanced Industrial Science and Technology (AIST), NMIJ, Tsukuba 305-8568, Japan

E-mail: fukuzawa@tagen.tohoku.ac.jp

Abstract. We have investigated charge separation Ne^+-Ne^+ and Ne^{2+}-Ne^+ from Ne_2 after direct double photoionization and compared them with those after 1s photoionization followed by KLL Auger decay and also with those after 2p photoionization of Ar_2 followed by LMM Auger decay. The Ne^+-Ne^+ separation takes place via radiative charge transfer from one-site two-hole states $Ne^{2+}(2p^4)$-Ne, whereas the Ne^{2+}-Ne^+ separation takes place via interatomic Coulombic decay from one-site two-hole states $Ne^{2+}(2s^12p^5)$-Ne.

1. Introduction

Both direct double photoionization and core-level photoionization followed by the Auger decay of molecules usually lead to ion-pair formation. The situation may be different for van der Waals rare gas dimers such as Ne_2. Both direct double photoionization and 1s photoionization of Ne_2 followed by KLL Auger decay create two holes in one Ne site. Such dicationic states are not necessarily dissociative [1]. In the present study, we will investigate the fate of such one-site two-hole states of Ne_2 created by direct double photoionization as well as 1s photoionization followed by KLL Auger decay. For comparison, we will also present results for the one-site two-hole states of Ar_2 created by 2p photoionization followed by LMM Auger decay.

For Ar_2, Morishita *et al* [2] found that $Ar^{2+}(3p^33d)$-Ar one-site state is subject to interatomic Coulombic decay (ICD) to the $Ar^{2+}(3p^4)$-Ar^+ charge-separation channel, using electron-ion multi-coincidence momentum spectroscopy (often called COLTRIMS or reaction microscope). Saito *et al* [3], on the other hand, elucidated that Ar^+-Ar^+ charge separation takes place from the $Ar^{2+}(3p^4)$-Ar one-site two-hole states, using Auger-electron-Ar^+-Ar^+ coincidence methods. Considering that the $Ar^{2+}(3p^4)$-Ar states are non-dissociative [4], it is concluded that the only possible mechanism of this charge separation is via radiative charge transfer. For Ne_2, Kreidi *et al* [5] and Yamazaki *et al* [6] have investigated 1s photoionization of Ne_2 using electron-ion multi-coincidence momentum spectroscopy and found that $Ne^{2+}(2p^4)$-Ne^+ charge separation takes place via ICD from one-site state $Ne^{2+}(2s^12p^5)$-

Ne populated by the $KL_1L_{2,3}$ Auger decay. Kreidi *et al* [7] did further extensive study on charge separation channels that include ICD, radiative charge transfer, and potential curve crossing from many different KLL Auger final states in Ne_2. In the present paper, we report the first observations of charge-separation *via* radiative charge transfer and *via* ICD, after direct double photoionization of Ne_2, and compare them with those after KLL Auger decay of Ne_2 as well as those after LMM Auger decay of Ar_2.

2. Experiment

The experiment was carried out on the c branch of the soft X-ray photochemistry beam line 27SU at SPring-8 [8-10]. The operation mode of the storage ring was the so called several single-bunches mode, i.e., 26 single-bunches + 2/29 filling mode, with a single-bunch separation of 165.2 ns.

The cluster beam is produced by expanding a mixture of neon and argon gases at a flow rate ratio of 70:1, at a total stagnation pressure of 1.2×10^5 Pa at 103 K, through a nozzle with diameter of 80 μm. Under these conditions, the cluster beam includes Ne and Ar monomers, Ne_2 and Ar_2 dimers, NeAr hetero-dimers, as well as larger clusters. The cluster beam is directed vertically.

The photon beam was focused to a size of less than 0.2 mm in height and 0.5 mm in width at the point of crossing with the cluster beam. The coincidence measurements described below were performed with the electric vector E of the linearly polarized light orientated vertically, at a photon energy of 268.24 eV, i.e., 19.61 and 17.46 eV above the atomic Ar $2p^{-1}\,{}^2P_{3/2}$ and $2p^{-1}\,{}^2P_{1/2}$ ionization thresholds, i.e., 248.628 and 250.776 eV, respectively [11], and at photon energy of 888.09 eV, i.e., 17.88 eV above the atomic Ne $1s^{-1}$ ionization thresholds, i.e., 870.21 eV [12]. The photon bandwidth is 0.1 eV and 0.6 eV at 268.24 eV and 888.09 eV, respectively.

Our momentum-resolved electron-ion-ion coincidence spectroscopy [13-19] is based on recording the electron and ion times-of-flight (TOFs) with multi-hit two-dimensional position sensitive detectors [20]. Knowledge of position and arrival time on the particle detectors, (x, y, t), allows us to extract information about the linear momentum (p_x, p_y, p_z) for each particle. The two TOF spectrometers are placed face to face. The TOF spectrometer axis is horizontal and perpendicular to both the photon beam and the molecular beam. The lengths of the acceleration region and the drift region of the electron spectrometer are 33.7 mm and 67.4 mm, respectively. For the ion spectrometer, there are two acceleration regions and no drift region. The length of the first acceleration region is 16.5 mm and that of the second one is 82.5 mm. The TOF spectrometer for the electron is equipped with a hexagonal multi-hit position-sensitive delay-line detector of effective diameter of 120 mm, while that for the ion is of effective diameter of 80 mm.

In the present experiments, the static extraction field was set to 1.5 V/mm. The static field of the second acceleration region for the ions was set to 21 V/mm. A uniform magnetic field of 6.1 G was superimposed to the spectrometer by a set of Helmholtz coils outside the vacuum chamber. Under these conditions, all the electrons up to the kinetic energies of 20 eV and all the ions up to the kinetic energies of 9 eV, both ejected in 4π sr, were accelerated onto the MCP detectors. The TOFs of the electrons and ions were recorded with respect to the bunch marker of the synchrotron radiation source using multi-hit time-to-digital converters (Roentdek TDC8HP). Appropriate gates selected only those electron signals synchronized with the single bunches. We recorded only events in which at least one ion and one electron were detected in coincidence.

3. Results and discussion

3.1. Photoion-photoion coincidence (PIPICO) map

Figure 1 shows the PIPICO map, simultaneously recorded for mixture gas of Ne and Ar (1.4 %) recorded at 268.24 eV. In this map, the horizontal and vertical axes are the TOFs for the first-arrived and second-arrived ions of the detected pairs. Only the Ar^+-Ar^+, Ar^+-Ar^{2+}. Ne^+-Ne^+ and Ne^+-Ne^{2+} pairs satisfying the momentum conservation laws within the plane perpendicular to the TOF axis are plotted.

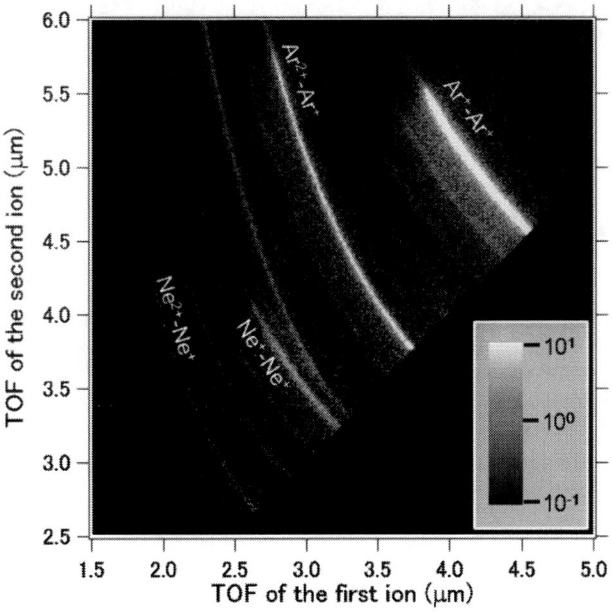

Figure 1. PIPICO map with Ne and Ar gas mixture, recorded at 268.24 eV.

Other ion pairs arising from hetero-dimers NeAr were also detected and will be discussed elsewhere. Some false coincident events originating from the previous bunch shifted by 165.2 ns are still seen. The results shown below are after imposing also the momentum conservation parallel to the TOF axis and thus all the false coincidences in figure 1 are further suppressed. Coincidence events originating from the ^{22}Ne isotopes are also seen in figure 1 but were not included in the following analysis.

3.2. Ar^+-Ar^+ coincident channel

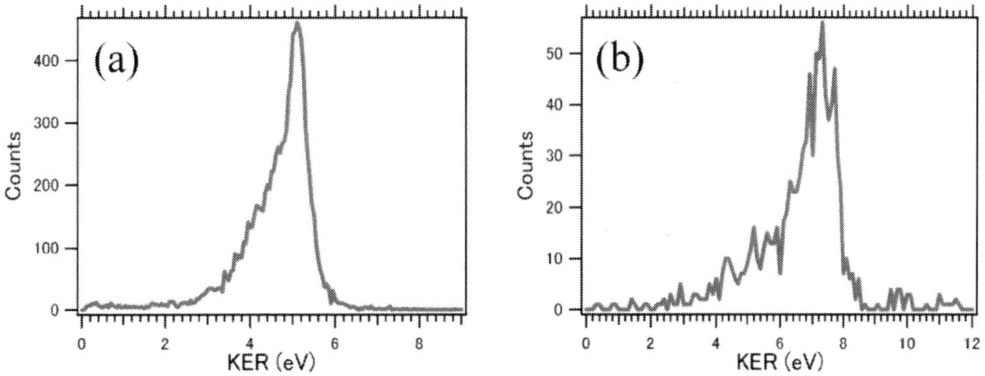

Figure 2. (a) Kinetic energy release distribution for Ar^+-Ar^+ pair recorded in coincidence with Ar 2p photoelectrons at the photon energy of 268.24 eV. (b) Kinetic energy release distribution for Ne^+-Ne^+ pair recorded in coincidence with Ne 1s photoelectrons at the photon energy of 888.09 eV.

Let us start with the strongest channel in the map of figure 1, i.e., Ar^+-Ar^+ pair formation. This is the channel intensively discussed [3]. In brief, figure 2 (a) depicts distribution of kinetic energy release (KER), or the sum of the kinetic energies of the two Ar^+ ions, recorded in coincidence with the Ar 2p photoelectrons at kinetic energy 18 eV. The KER is peaked at 5.1 eV. Assuming a pure Coulomb repulsive energy between the two Ar^+ ions, one can estimate the internuclear distance at the Coulomb explosion to be 2.8 Å. This internuclear distance is much shorter than equilibrium internuclear distance, 3.8 Å, in the neutral Ar_2 dimers. On the other hand, it is in good agreement with the equilibrium internuclear distances of the one-site dicationic states $Ar^{2+}(3p^4\ {}^3P,\ {}^1D,\ {}^1S)$-Ar [4]. These states can be populated via the intra-atomic $L_{2,3}M_{2,3}M_{2,3}$ Auger decay in one site, following Ar 2p ionization. The sum of the branching ratios to these Auger final states is 0.72. These Auger-final states are, however, not dissociative. The only way to produce Ar^+-Ar^+ pair from these Auger-final states is radiative charge transfer, as discussed previously.

3.3. Ne^+-Ne^+ coincident channel

PIPICO map recorded at 888.09 eV, not shown here, exhibits the strong Ne^+-Ne^+ coincidence channel. KER distribution recorded in coincidence with Ne 1s photoelectrons at kinetic energy of 18 eV shown in fugure 2 (b) is peaked at 7.4 eV, illustrating that the charge separation starts to take place at the internuclear distance of 1.9 Å. This internuclear distance is much shorter than the equilibrium internuclear distance, 3.1 Å, in the neutral Ne_2 dimers but coincides with the equilibrium internuclear distances of one-site dicationic states $Ne^{2+}(3p^4\ {}^3P,\ {}^1D,\ {}^1S)$-Ne [1]. Thus, as in the case of Ar_2 discussed above, Ne^+-Ne^+ pair formation is mostly via radiative charge transfer from these one-site states

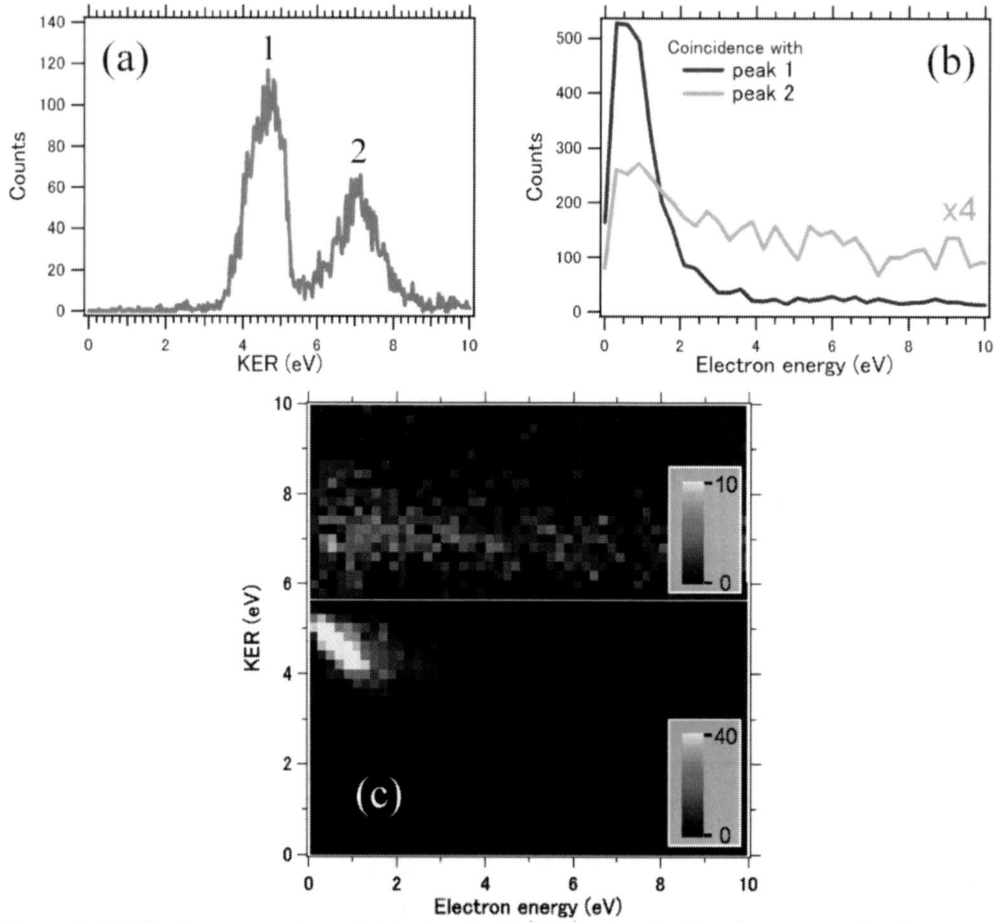

Figure 3. (a) Kinetic energy release distribution for Ne^+-Ne^+ pair coincident formation at the photon energy of 268.24 eV. (b) Electron kinetic energy spectrum coincident with Ne^+-Ne^+ pair at the photon energy of 268.24 eV. (c) Relationship between the electron energy and the KER.

produced via intra-atomic Ne $KL_{2,3}L_{2,3}$ Auger decay. The sum of the branching ratio to these Auger-final states is 0.70. This channel was observed also by Kreidi *et al* [5,7] and Yamazaki *et al* [6].

Let us now discuss the Ne^+-Ne^+ pair formation at 268.24 eV. Figure 3 (a) depicts KER distributions. Two peaks appear: peak 1 is at 4.8 eV and peak 2 is at 7.1 eV. Assuming a pure Coulomb repulsive energy between the two Ne^+ ions, one can estimated the internuclear distances 3.0 Å and 2.0 Å for the peaks 1 and 2, respectively. Figure 3 (b) depicts two electron energy spectra coincident with the Ne^+-Ne^+ pairs that form the peaks 1 and 2, respectively. The electron spectrum that correlates to peak 1 is peaked at 0.4 eV. Both KER 4.8 eV and the electron energy 0.4 eV well agree with the measurement by Jahnke *et al* [21], where they identified the interatomic Coulombic decay in the Ne dimer after Ne 2s ionization. The relationship between the electron energy and the KER is shown in figure 3 (c). The tilted island with a slope of -1 around the KER of 4.8 eV and the electron energy of 0.4 eV is providing evidence that Ne^+-Ne^+ pair formation is via ICD decay following 2s ionization. The energy of the peak 2 coincides with that in figure 2 (b). Furthermore, the corresponding electron spectrum indicates that most of the contributions come from continuous electron emission. These facts indicate that the channels forming the peak 2 with continuous electron emission are mostly attributed to radiative charge transfer after 2p double photoionization. In other words, this is the first evidence of radiative charge transfer that takes place after double photoionization. We note, however, that the energy of the peak 2 also coincides with that of shake-induced ICD observed by Jahnke *et al* [21]. The intensity of the peak 2 observed in the present experiment is much higher than they observed. Thus, to be exact, the present peak 2 is considered to be the sum of radiative charge transfer from the $2p^4$ dicationic states the shake-up induced ICD from $2p^4np$ cationic satellite states

3.4. Ar^{2+}-Ar^+ and Ne^{2+}-Ne^+ coincident channels

In figure 4, the panels (a), (b), and (c) depict the KER distributions for Ar^{2+}-Ar^+ formation at 268.24 eV, Ne^{2+}-Ne^+ formation at 888.09 eV, and Ne^{2+}-Ne^+ formation at 268.24 eV, respectively, whereas panels (d), (e), and (f) show the electron energy distributions (< 10 eV) recorded in coincidence with Ar^{2+}-Ar^+ at 268.24 eV, Ne^{2+}-Ne^+ at 888.09 eV, and Ne^{2+}-Ne^+ at 268.24 eV, respectively. As can be seen, low-energy electron yields are common for all these three cases.

Let us first focus on the spectra (a) and (d) for Ar^{2+}-Ar^+ formation at 268.24 eV. This channel was first discussed by Morishita *et al* [2]. Ar^{2+}-Ar^+ formation is mostly due to interatomic Coulombic decay from the Auger final one-site dicationic state $Ar^{2+}(3p^33d)$-Ar to the two-site tri-cationic state $Ar^{2+}(3p^4)$-$Ar^+(3p^5)$. The low-energy electrons (~2 eV) in the panel (d) is emitted by this ICD process. The KER distribution in the panel (a) is peaked at 7.4 eV, illustrating that the charge separation starts to take place at the internuclear distance of 3.9 Å, i.e., close to the equilibrium internuclear distance 3.8 Å, and thus that the ICD and Auger take much faster than nuclear motion.

Let us focus on the spectra (b) and (e) for Ne^{2+}-Ne^+ at 888.09 eV. This channel was again fully discussed by Kreidi *et al* [5,7] and by Yamazaki *et al* [6]. Ne^{2+}-Ne^+ formation is mostly due to interatomic Coulombic decay from the Auger final one-site dicationic state $Ne^{2+}(2s^12p^5\,{}^1P)$-Ne to the two-site tri-cationic states $Ne^{2+}(2p^4\,{}^1D)$-$Ne^+(2p^5)$ and $Ne^{2+}(2p^4\,{}^3P)$-$Ne^+(2p^5)$. In the KER distribution in the panel (b), the larger peak at 9.2 eV correspond to the channel to $Ne^{2+}(2p^4\,{}^1D)$-$Ne^+(2p^5)$, whereas the smaller peak at 12.5 eV to $Ne^{2+}(2p^4\,{}^3P)$-$Ne^+(2p^5)$. The former peak corresponds to the case when the charge separation starts to take place at the equilibrium internuclear distance 3.1 Å of the neutral Ne_2 ground state, whereas the latter peak to the case when the charge separation starts to take place at shortened internuclear distance 2.3 Å. In the first case, the spin of the double charged Ne atom is conserved, whereas the spin is not conserved in the second case. These differences come from the different ICD mechanisms. The first type of ICD is caused by the direct Coulomb integral and thus called direct ICD. The second time of ICD is caused by the exchange integral and thus called exchange ICD. Direct ICD dominates at large interculear distances, whereas exchange ICD becomes effective at short internuclear distances where the wavefunctions overlap. In the present Ne_2 case, the direct ICD is energetically closed at the interculear distances < 2.5 Å. Thus we see clearly the

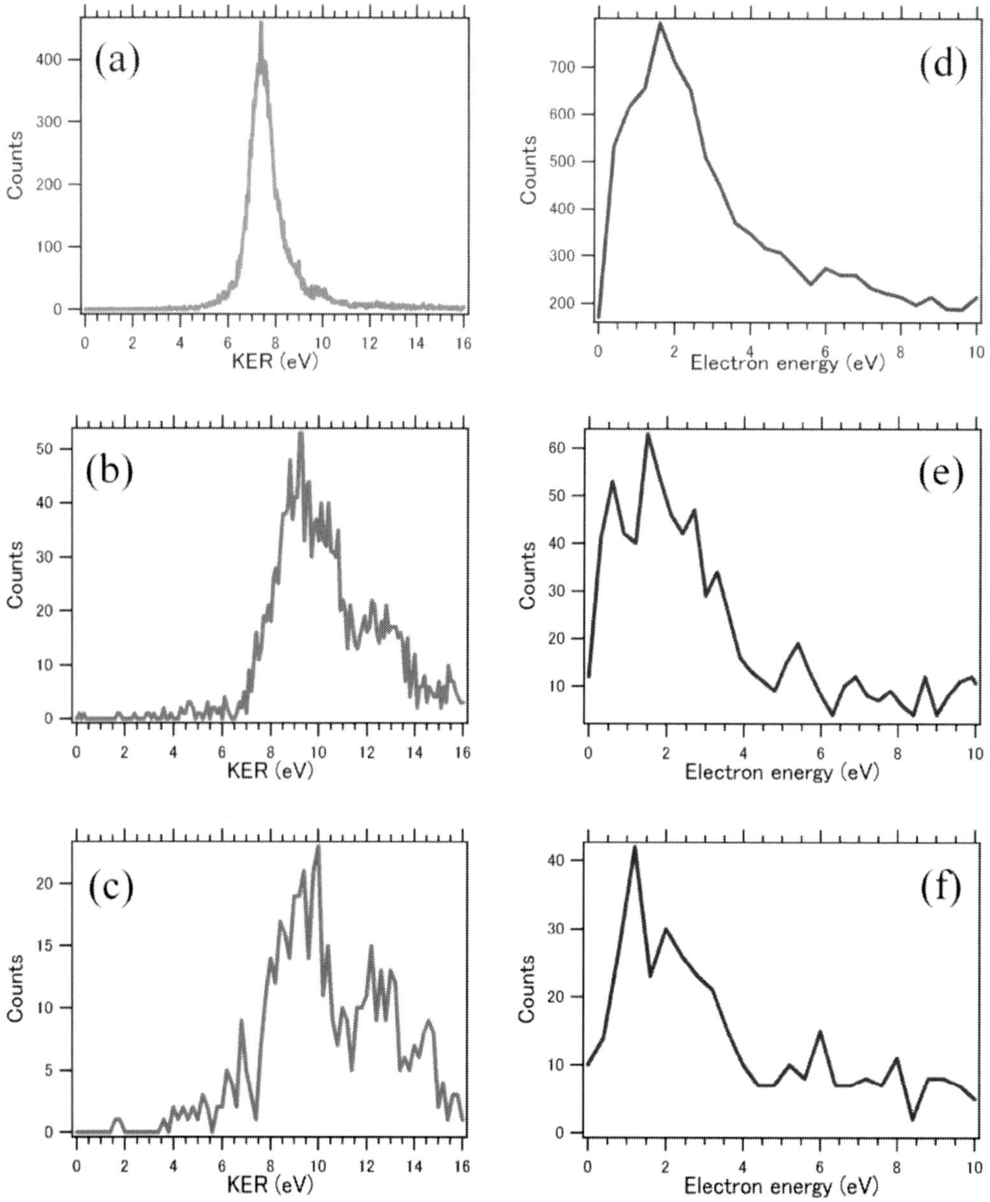

Figure 4. (a) Kinetic energy release distribution for Ar^{2+}-Ar^{+} at 268.24 eV. (b) Kinetic energy release distribution for Ne^{2+}-Ne^{+} at 888.09 eV. (c) Kinetic energy release distribution for Ne^{2+}-Ne^{+} at 268.24 eV. (d) Electron energy distribution for Ar^{2+}-Ar^{+} at 268.24 eV. (e) Electron energy distribution for Ne^{2+}-Ne^{+} at 888.09 eV. (f) Electron energy distribution for Ne^{2+}-Ne^{+} at 268.24 eV.

exchange ICD at internuclear distances < 2.5 Å. The low-energy electrons (~2 eV) in the panel (e) are emitted by these two types of ICD processes.

Now let us focus on the spectra (c) and (f) for Ne^{2+}-Ne^{+} at 268.24 eV. We immediately notice that these spectra well coincide with spectra (b) and (e) just discussed above. Thus the spectra (c) and (f) is considered to be due to ICD from one-site dicationic state $Ne^{2+}(2s^12p^5\ ^1P)$-Ne. The only possible mechanism to produce one-site dicationic state $Ne^{2+}(2s^12p^5\ ^1P)$-Ne at photon energy 268.24 eV is double photoionization that directly creates $Ne^{2+}(2s^12p^5\ ^1P)$-Ne. Thus this is the first evidence of ICD that takes place after double photoionization.

4. Conclusion

We have observed charge separations Ne^+-Ne^+ and Ne^{2+}-Ne^+ from Ne_2 after direct double photoionization and after 1s photoionization and compared them with charge separations Ar^+-Ar^+ and Ar^{2+}-Ar^+ from Ar_2 after 2p photoionization, using electron-ion-ion coincidence momentum spectroscopy. We found that the charge separation Ne^+-Ne^+ takes place via radiative charge transfer from one-site two-hole states $Ne^{2+}(2p^4)$-Ne populated via the direct double photoionization as well as via $KL_{2,3}L_{2,3}$ Auger decay following 1s photoionization, whereas the charge separation Ne^{2+}-Ne^+ takes place via ICD from one-site two-hole states $Ne^{2+}(2s^12p^5)$-Ne populated via the direct double photoionization as well as via $KL_1L_{2,3}$ Auger decay following 1s photoionization. These are the first evidences of the radiative charge transfer and ICD after direct double photoionization. Measurements of the radiative charge transfer and ICD following direct double ionization of Ar_2 at the photon energy below Ar 2p ionization threshold would be possible.

Acknowledgements

The experiments were carried out at SPring-8 with approval by JASRI. We are grateful to Lorenz Cederbaum and Alexander Kueff for helpful discussion. The work was partially supported by Grants-in-Aid for Scientific Researches from Japan Society for the Promotion of Science (JSPS) and by the Budget for Nuclear Research from Ministry of Education, Culture, Sports, Science and Technology, based on screening and counseling by the Atomic Energy Commission. DD and DZ thank the hospitality during their staying in SPring-8 and the partial support from NSFC.

References

[1] Stoychev S D, Kuleff A I, Tarantelli F and Cederbaum L S 2008 *J. Chem. Phys.* **129** 074307
[2] Morishita Y *et al* 2006 *Phys. Rev. Lett.* **96** 243402
[3] Saito N *et al* 2007 *Chem. Phys. Lett.* **441** 16
[4] Stoychev S D, Kuleff A I, Tarantelli F and Cederbaum L S 2008 *J. Chem. Phys.* **128** 014307
[5] Kreidi K *et al* 2008 *J. Phys. B: At. Mol. Opt. Phys.* **41** 101002
[6] Yamazaki M *et al* 2008 *Phys. Rev. Lett.* **101** 043004
[7] Kreidi K *et al* 2008 *Phys. Rev.* A **78** 043422
[8] Ohashi H *et al* 2001 *Nucl. Instrum. Methods Phys. Res.* A **467-468** 529
[9] Ohashi H *et al* 2001 *Nucl. Instrum. Methods Phys. Res.* A **467-468** 533
[10] Ueda K 2003 *J. Phys. B: At. Mol. Opt. Phys.* **36** R1
[11] King G C *et al* 1977 *J. Phys. B: At. Mol. Phys.* **10** 2479
[12] Siegbahn K *et al* 1969 *ESCA Applied to Free Molecules* (Amsterdam, North-Holland)
[13] Muramatsu Y *et al* 2002 *Phys. Rev. Lett.* **88** 133002
[14] Fanis A De *et al* 2001 *Phys. Rev. Lett.* **89** 023006
[15] Ueda K and Eland J H D 2005 *J. Phys. B: At. Mol. Opt. Phys.* **38** S839
[16] Saito N, Fanis A De, Koyano I and Ueda K 2005 *J. Electron Spectrosc. Relat. Phenom.* **144-147** 103
[17] Morishita Y *et al* 2006 *Radiat. Phys. Chem.* **75** 1977
[18] Ueda K *et al* 2007 *J. Electron Spectrosc. Relat. Phenom.* **155** 113
[19] Saito N, Liu X-J, Morishita Y, Suzuki I H and Ueda K 2007 *J. Electron Spectrosc. Relat. Phenom.* **156–158** 68
[20] Jagutzki O *et al* 2002 *IEEE Transact. Nucl. Science* **49** 2477
[21] Jahnke T *et al* 2004 *Phys. Rev. Lett.* **93** 163401

Vibrationally resolved resonant x-ray emission spectra of diatomic molecules

M Oura[1], O Takahashi[2], T Gejo[1,3], T Tokushima[1], Y Horikawa[1], Y Senba[4], H Ohashi[1,4], S Shin[1,5]

[1]RIKEN SPring-8 Center, 1-1-1 Kouto, Sayo-cho, Sayo-gun, Hyogo 679-5148, Japan
[2]Department of Chemistry, Hiroshima University, Higashi-Hiroshima 739-8526, Japan
[3]Dep. of Material Science, University of Hyogo, Ako-gun, Hyogo 678-1297, Japan
[4]JASRI/SPring-8, 1-1-1 Kouto, Sayo-cho, Sayo-gun, Hyogo 679-5198, Japan
[5]Institute for Solid State Physics, The University of Tokyo, Kashiwa, Chiba 277-8581, Japan

E-mail: `oura@spring8.or.jp`

Abstract. We report here the measurements of vibrationally resolved resonant x-ray emission spectra of core-excited N_2 and O_2 molecules in the $1s \rightarrow \pi^*$ resonance region. The measured spectra were compared with the simulated spectra calculated based on the lifetime-vibrational interference theory. We find that the specific excitation to the $v' = 0$ vibrational level in the intermediate core-hole state of N_2 molecule leads to clear vibrational structure ($v'' = 0-3$) in both of the final electronic X $^1\Sigma_g^+$ (participator band) and a $^1\Pi_g$ (spectator band) states. In contrast to that, the excitation to higher-v' vibrational levels gives rise to the complicated structure in the resultant soft x-ray emission spectra. In the case of O_2 molecule, we have observed some structures arising from the vibrational progression, but we find that much higher resolution for the spectrometer is needed to observe distinct vibrational levels.

1. Introduction

Soft x-ray emission (SXE) spectroscopy with the aid of energy selective excitation provides an excellent tool for investigating the electronic structure and determining symmetries of occupied and unoccupied orbitals in molecules [1, 2, 3, 4, 5, 6], since the dipole nature restricts the possible symmetries of all the states involved in the absorption / emission processes. With the advent of synchrotron radiation (SR) sources, the resonant SXE spectroscopy has been successfully applied to diatomic molecules, such as N_2 [3], O_2 [4] and CO [5, 7], as they provide us with simple but excellent test systems to advance the basic understanding of the electronic structure as well as the dynamic processes concerned.

Recent advancement of the SXE spectrometer, e.g., throughput and resolution, will provide us with an opportunity to gain further knowledge of the excitation / relaxation processes in core-excited molecules. Recently, Tokushima *et al* developed a high performance slit-less spectrometer for SXE spectroscopy [8] which is applicable to gas-phase experiment. By taking this occasion, we have applied the resonant SXE spectroscopy to diatomic N_2 and O_2 molecules in the region of $1s \rightarrow \pi^*$ resonance to study the effect of nuclear motion on the excitation / relaxation processes.

In this article, we report the measurements of vibrationally resolved resonant SXE spectra of core-excited N_2 and O_2 molecules. Although these molecules have been previously studied by means of the resonant SXE spectroscopy [3, 4], we show the spectra measured with the resolving power $E/\Delta E$ of $\sim 1{,}950$ for N_2 and of $\sim 2{,}040$ for O_2, which are better than those of previous work ($E/\Delta E \sim 790$ for N_2 [3] and ~ 530 for O_2 [4]), in the present study. It is obvious that the present spectra are of higher

Journal of Physics: Conference Series **235** (2010) 012016 doi:10.1088/1742-6596/235/1/012016

quality to investigate the vibrational structure in the resonant SXE spectra. We compare the measured spectra with the simulations based on the lifetime-vibrational interference theory.

2. Experimental

The experiments were carried out on the a-branch of the soft x-ray undulator beamline BL17SU at SPring-8, an 8 GeV SR facility in Japan. This beamline is equipped with the high-resolution varied-line-spacing-plane-grating monochromators [9], which can provide us with highly stabilized monochromatic soft x rays with resolving power of over 10,000 [10]. The direction of the polarization vector, e.g., horizontal or vertical, of the incident photon beam, generated by the multi-polarization-mode undulator [11, 12, 13], can be switched by changing the operational mode of the undulator when the photon energy is higher than 500 eV [14]. Although the degree of linear polarization is not so high, e.g., $P_L \sim 0.82$ (-0.73) at 500 eV, we can perform the angle-resolved SXE spectroscopy without rotating the spectrometer. In the present study, however, we have set up the multi-polarization-mode undulator to be a helical mode, e.g., $P_C \sim 0.98$ (0.95) at 530 eV (400 eV), so as to achieve higher photon flux.

Energy calibration of the incident photon beam was performed by recording the total ion yield spectra of N_2 and O_2 molecular gases in the inner-shell excitation region, and the absolute energy scale was established using the reference energies published in the previous paper [15, 16]. The accuracy of energy scale for the incident photon beam was estimated to be about ± 40 meV. Although the monochromator of BL17SU has ability to achieve the resolving power $E/\Delta E > 15{,}000$, the photon band pass, at a photon energy of 530 eV (400 eV), was set to be about 65 meV (39 meV) to achieve higher photon flux in the present study.

The x-ray spectra emitted from the core-excited molecules were recorded by using the high performance slit-less spectrometer equipped with the compact flange-mounted liquid flow-cell [8, 17]. Although this spectrometer was originally aimed at performing the spectroscopic studies on the liquid target [18, 19], in the present study we have introduced the target molecular gas into the flow-cell instead of the liquid. The flow-cell utilizes a 150 nm thick Au-coated SiC membrane window to separate the flowing gas at atmospheric pressure from the high vacuum. Calibration of the SXE spectrometer was performed by measuring the normal fluorescence emitted from the core-ionized molecules and also the photons elastically scattered from the target gas. The estimated energy resolution of the SXE spectrometer was 258 meV (203 meV) for the experiment on the O_2 (N_2) molecule.

3. Results and discussion

3.1. Total ion yield spectra

Figure 1 shows the total ion yield spectra of molecular (a) N_2 and (b) O_2 in the region of $1s \rightarrow \pi^*$ excited states. Both spectra were fitted by means of a least squares fitting procedure employing Voigt functions for the vibrational progression, where thin solid curves indicate the individual components of the vibrational progression and thick curves are the sum of them. Lorentzian width (ω_L) and Gaussian width (ω_G) were assumed to be the same for the individual components in the fitting procedure.

For the N_2 molecule, shown in Fig.1(a), we have tried to fit the observed spectrum by adopting the spectroscopic constants reported by Feifel *et al* [20] for the vibrational energy levels of the N_2 (N $1s, v = 0) \rightarrow (\pi^*, v' = 0-6)$ transitions. We obtained $\omega_L = 109$ meV and $\omega_G = 39$ meV from the present fitting procedure. The obtained value of 109 meV for ω_L is slightly smaller than the recently reported values, which are ranging from 113 meV to 116 meV [16, 21, 22, 23], but identical to the value reported by Ohashi *et al* [24] where the fitting was applied to the total ion yield spectrum. For the O_2 molecule, shown in Fig.1(b), we have adopted the actual lifetime width of the core-excited state ($\Gamma = 149$ meV) for ω_L and the spectroscopic constants ω_e and $\omega_e\chi_e$ [25] for the vibrational energy levels of the O_2 (O $1s, v = 0) \rightarrow (\pi^*, v' = 0-n)$ transitions. In the present analysis, we have assumed the vibrational progressions up to $n = 15$. As shown in Fig.1, the results of these fitting demonstrate to be good enough to validate the employed spectroscopic constants for the intermediate core-hole state.

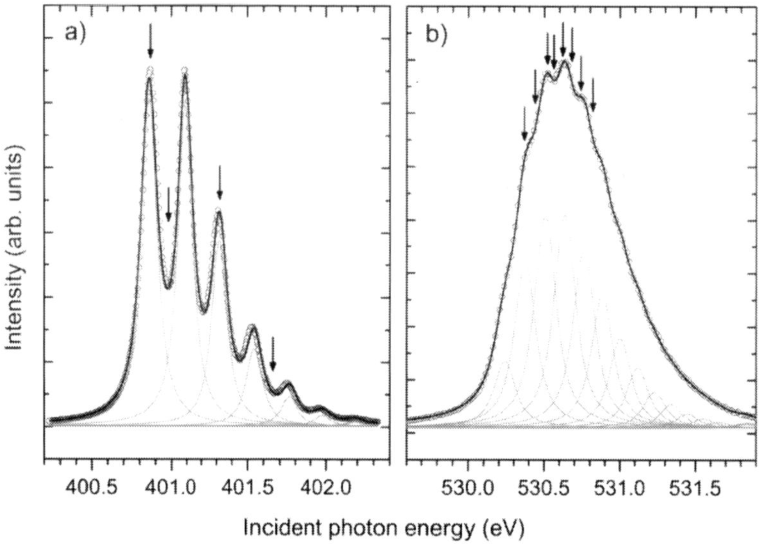

Figure 1. Total ion yield spectra of molecular (a) N_2 and (b) O_2 in the region of $1s \to \pi^*$ excited states. Solid curves depict the results of the fitting ; thin solid curves are the individual components represented as Voigt functions and thick curves are the sum of them. Arrows indicate the photon energies where the SXE spectra were measured.

Arrows in Fig.1 indicate the photon energies at which the SXE spectra were measured. As can be seen in the total ion yield spectra, the vibrational structure for N_2 molecule is more clearly resolved as compared to the case of O_2 molecule. In both cases, however, the lifetime width associated with core-hole state is nearly the same as the vibrational spacing. This implies an excitation with specific energy may cause a coherent excitation to more than one vibrational level due to the fact that the core-hole lifetime width is approximately same as the vibrational spacing in the excited state.

3.2. Soft x-ray emission spectra

Figures 2 and 3 show the SXE spectra of N_2 and O_2 molecules, respectively. All the SXE spectra shown in Figs 2 and 3 are raw spectra, i.e., no correction is made for self-absorption and absorption by the Au-coated SiC membrane window. All the spectra have exhibited a participator band as well as a spectator band. In the case of the participator decay, the excited electron fills the core-hole giving rise to the final electronic state identical to the ground state of the molecule. In the case of the spectator decay, on the other hand, one of the valence electrons, i.e., $3\sigma_g$ orbital, fills the inner-shell vacancy while the excited electron remains in the π_g orbital. Electrons in the inner-valence $2\sigma_u$ and $1\pi_u$ orbitals are not allowed to fill the core-hole owing to the parity selection rule. For both spectra of N_2 and O_2 molecules, we could clearly recognize the fine structure, i.e., vibrational structure. This fine structure has not been clearly observed in the previous experiments [3, 4].

The SXE spectrum of N_2 molecule measured at 401.32 eV is represented together with the results of Gaussian profile fitting. Short vertical bars in Figs 2 and 3 indicate calculated vibrational energies for the final electronic X $^1\Sigma_g^+$ (participator band) and a $^1\Pi_g$ (spectator band) states using the spectroscopic constants available from the literature [26, 28]. It is unfortunate that no reliable spectroscopic constants

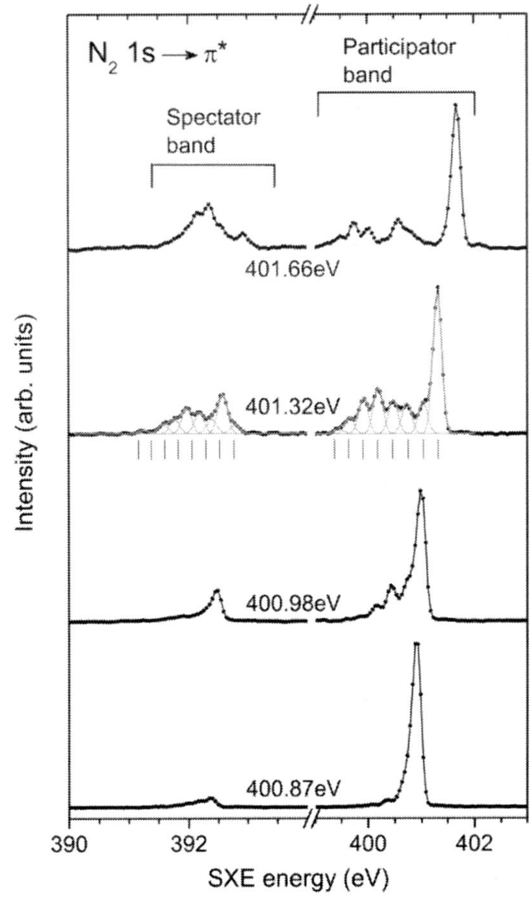

Figure 2. Soft x-ray emission (SXE) spectra of N_2 molecule recorded for four different excitation energies. No correction was made for self-absorption and absorption by the Au-coated SiC membrane window. The SXE spectrum measured at 401.32 eV is represented together with the results of Gaussian profile fitting. Short vertical bars in the figure indicate calculated vibrational energies for the final electronic X $^1\Sigma_g^+$ (participator band) and a $^1\Pi_g$ (spectator band) states using the spectroscopic constants [28].

for the O_2 spectator band are available from the literature. For the N_2 spectrum, especially for the participator band, we can recognize that the calculated energy positions agree well with the center positions of each Gaussian profile. As can be seen in these figures, the observed fine structure can be explained as the vibrational progressions. Again we can notice that the vibrational structure for N_2 molecule is more obvious than that of O_2 molecule. Thus the SXE spectrometer with much higher resolution is indispensable to study the vibrational structure more clearer for O_2 molecule.

3.3. Absorption corrections

In order to interpret the observed SXE spectra correctly, we stress here that the corrections for the spectrum, especially due to the self-absorption correction, is indispensable, since the target gas in the

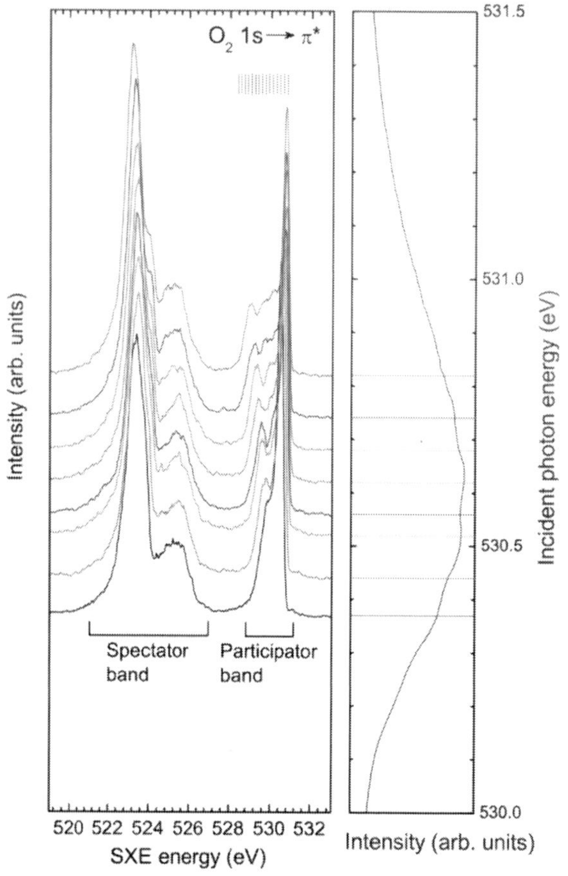

Figure 3. Soft x-ray emission (SXE) spectra of O_2 molecule (left panel) recorded by tuning the excitation energies to the energy points indicated by arrows in Fig.1(b). The corresponding energies are also shown as solid lines in the total ion yield spectrum (right panel). In the SXE spectra, no correction was made for self-absorption and absorption by the Au-coated SiC membrane window. Short vertical bars in the figure depict calculated vibrational energies for the final electronic $X^3\Sigma_g^+$ (participator band) states using the spectroscopic constants [26].

flow-cell is in atmospheric pressure and the energy region of interest overlays with the π^* resonance. For the proper absorption corrections, the following steps are needed to proceed ;

(i) absorption correction for the incoming SR beam by the Au-coated SiC membrane window and the target gas itself,
(ii) self-absorption correction for the outgoing SXE spectrum by the target gas,
(iii) absorption correction for the outgoing SXE spectrum by the Au-coated SiC membrane window.

For the correction (ii), the volume of gas-cell along the incoming and outgoing paths were divided into thin layers, and we corrected the intensity distribution of the SXE spectrum in each thin layer. Now let us consider, for example, the correction for 4th thin layer. The intensities $I(E_x, E)$ of the incident, absorbed and measured flux for 4th thin layer can be described as :

$$I_4^{in}(E_x) = I_0 \exp\{-\mu_w(E_x)\rho_w x_w \mathrm{cosec}\,\theta_{in}\} \times \prod_{i=1}^{4-1} \exp\{-\mu_t(E_x)\rho_t x_t \mathrm{cosec}\,\theta_{in}\},$$

$$I_4^{abs}(E_x) = I_4^{in}(E_x)\langle 1 - \exp\{-\mu_t(E_x)\rho_t x_t \mathrm{cosec}\,\theta_{in}\}\rangle,$$

$$I_4^{spe}(E_x, E) \propto I_4^{abs}(E_x) S_4(E_x, E) \exp\{-\mu_w(E)\rho_w x_w \mathrm{cosec}\,\theta_{out}\} \times \prod_{i=1}^{4} \exp\{-\mu_t(E)\rho_t x_t \mathrm{cosec}\,\theta_{out}\},$$

where E_x is the excitation energy of in-coming radiation, I_0 the primary photon flux incident on the gas-cell, μ_w and μ_t the mass attenuation coefficients of the gas-cell window and the target [29, 16], ρ_w and ρ_t the densities of the gas-cell window and the target, x_w and x_t the thicknesses of the gas-cell window and the target thin layer, θ_{in} and θ_{out} the angles of the in-coming and out-going radiations with respect to the layer surface, E the energy of the out-going radiation to be measured, $S(E_x, E)$ the spectral function excited by the in-coming radiation, respectively. We assumed that the spectral function and the angular distributions of the photon-in–photon-out process for each thin layer are identical with each other.

Finally, we summed up the contribution from each thin layer to plausibly establish the absorption-corrected intensity distribution as a function of the SXE energy. Actually, the correction was performed to extract the spectral function $S(E_x, E)$ by using the spreadsheet. Although this simple procedure does not give us the absolute correction, it is good enough to overlook the relative energy dependence of the self-absorption correction factor. Figure 4 shows the typical example of absorption corrections made for the SXE spectrum of N_2 molecule measured at 401.66 eV excitation energy. In order to facilitate the comparison, the intensity of the corrected spectrum was normalized to that of the uncorrected one for the energy region of the spectator band. As we can recognize from the figure, the effect of the self-absorption correction is remarkable at the SXE energy region of 400.5 $\sim$ 401.7 eV, i.e., the energy of N_2 $1s \rightarrow \pi^*$ resonance region.

3.4. Spectral shape analysis

For the analysis of the spectral shape, we need to treat the problem beyond the ordinary Franck-Condon theory, because the coherently excited levels can cause interference effects where the levels can decay to the same final vibronic states. In such case, a conventional Franck-Condon analysis, in which the excitation and relaxation are treated as two independent processes, is no longer valid and a one-step formalism taking account of the lifetime-vibrational interference effects [30] is suitable to analyze the resonantly excited SXE spectrum. Similar spectral analyses have been made for the resonant Auger spectra of N_2 [31, 32] and O_2 [33] molecules and the resonant SXE spectra of N_2 [3] and CO [7] molecules. Accordingly, our simulation is based on the method similar to these previous studies.

The intensity distribution $I(E_x)$ of the SXE spectrum can be described as the summation of the population probability $I_f(E_x)$ on the same emission energy scale :

$$I(E_x) \propto \sum_f I_f(E_x).$$

In the present analysis, we calculated the $I_f(E_x)$ within the framework of the one-step formalism and also the two-step formalism. In the case of the one-step formalism, the $I_f(E_x)$ was calculated based on the lifetime-vibrational interference theory as

$$I_f(E_x) \propto G(E_x) \left| \sum_n \frac{< f|n >< n|0 >}{E_x - (E_n - E_0) + i\Gamma/2} \right|^2,$$

where E_x, E_n and E_0 are the excitation energy, the vibrational energy of the intermediate core-hole state and of the ground state. $G(E_x)$ is the monochromator function and Γ is the lifetime width of the

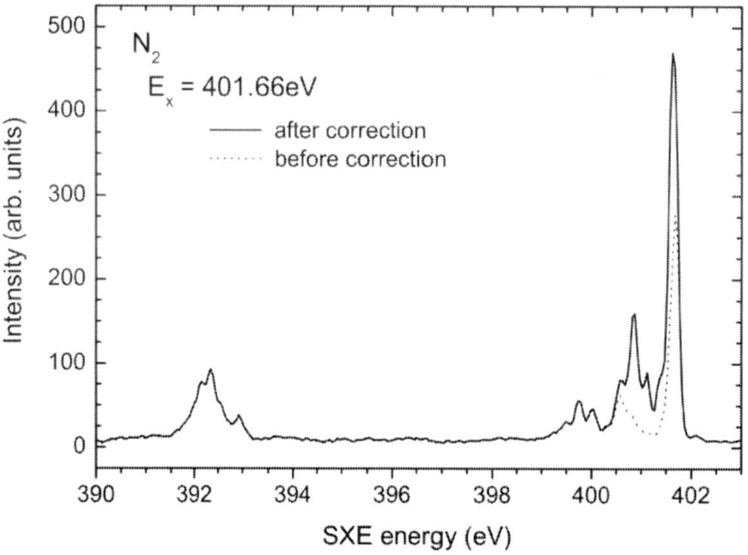

Figure 4. Soft x-ray emission (SXE) spectra of N_2 molecule measured at 401.66 eV. The thick solid line shows the spectrum with absorption corrections, whereas the dashed line represents the spectrum without any corrections.

intermediate core-hole state. By combining the above two formulae, we get the intensity distribution as :

$$I(E_x) \propto G(E_x)\sum_f\sum_n\left|\frac{<f|n><n|0>}{E_x-(E_n-E_0)+i\Gamma/2}\right|^2$$
$$+G(E_x)\sum_f\sum_{n\neq n'}\frac{<f|n><n|0><f|n'><n'|0>}{\{E_x-(E_n-E_0)+i\Gamma/2\}\{E_x-(E_{n'}-E_0)+i\Gamma/2\}}.$$

Here, the first term is called as *direct* term and the second term the *interference* term. In the case of the two-step formalism, on the other hand, the $I_f(E_x)$ was simply calculated by adopting the probability to produce the intermediate core-hole state $I_{XAS}(E_x)$ and the Franck-Condon factors between the intermediate state and the final state as

$$I_f(E_x) \propto \left|<f|n>\right|^2 \times I_{XAS}(E_x).$$

In the spectral shape analysis, we have used the well-established parameters listed in Table 1 for the potential curves of the ground, intermediate core-excited and final states [3, 20, 25, 27, 28]. We have confirmed the validities for some parameters as shown in Figs 1 and 2. The Franck-Condon factors, i.e., the squares of overlap integrals of the vibrational wave functions, were calculated using the Morse potentials. The intensity distribution of the SXE spectra were simulated by employing the one-step formalism based on the lifetime-vibrational interference theory [30]. For comparison, we also calculated the SXE spectra by means of the two-step formalism.

Figures 5 and 6 depict the results of the simulations for N_2 and O_2 molecules, respectively. In these figures, the solid circles represent the measured SXE spectra and the solid curves show the simulated

Table 1. Spectoscopic parameters of N_2 and O_2 molecules used for calculations of the Morse potential curves and the Franck-Condon Factors.

Configuration	state	T_{00} [a] /eV	ω_e [a] /cm^{-1}	$\omega_e x_e$ [a] /cm^{-1}	r_e [a] /Å
N_2 G.S.	$X\,^1\Sigma_g^+$	0.0	2358.57[b]	14.324[b]	1.09768[b]
NN* $1\sigma_u^{-1}1\pi_g$	$^1\Pi_u$	400.88[b]	1904.1[c]	17.235[c]	1.1645[d]
N_2 $3\sigma_g^{-1}1\pi_g$	$a\,^1\Pi_g$	8.55[e]	1694.2[e]	13.949[e]	1.2203[e]
O_2 G.S.	$X\,^3\Sigma_g^-$	0.0	1580.361[e]	12.073[e]	1.20739[e]
OO* $1\sigma_u^{-1}1\pi_g^3$	$^3\Pi_u$	530.24[f]	1116.271[g]	12.905[g]	1.3427[g]

[a] T_{00} is the excitation energy from the G.S. to the lowest ($v' = 0$) vibrational level of the excited state, ω_e the vibrational energy, $\omega_e x_e$ the anharmonicity constant, and r_e the equilibrium internuclear distance.
[b] [3].
[c] [20].
[d] [27]
[e] [28].
[f] Parameter determined by the present fitting procedure.
[g] [25].

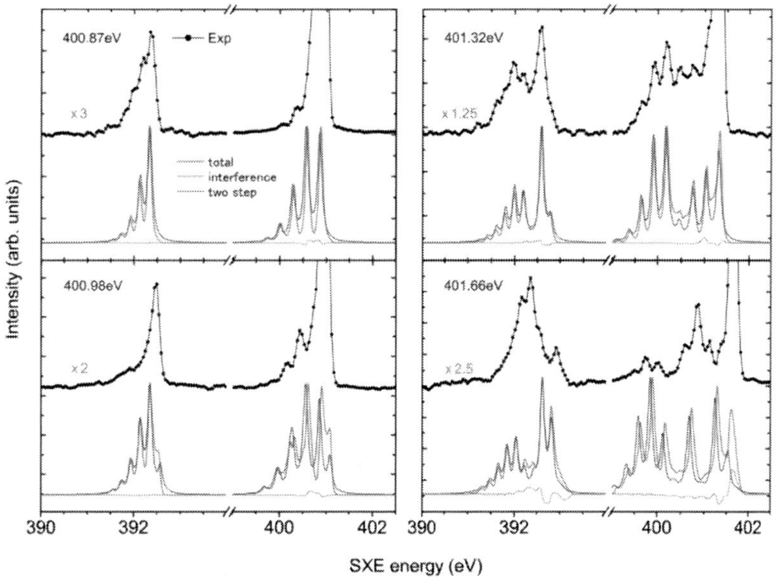

Figure 5. Comparison of the measured and simulated soft x-ray emission (SXE) spectra of N_2 molecule. Excitation energies for each spectrum are indicated near the spectra. Intensities in the spectator bands ranged between 390 and 394 eV are magnified for ease of comparison. The solid circles represent the measured spectra, while solid lines depict the simulated spectra.

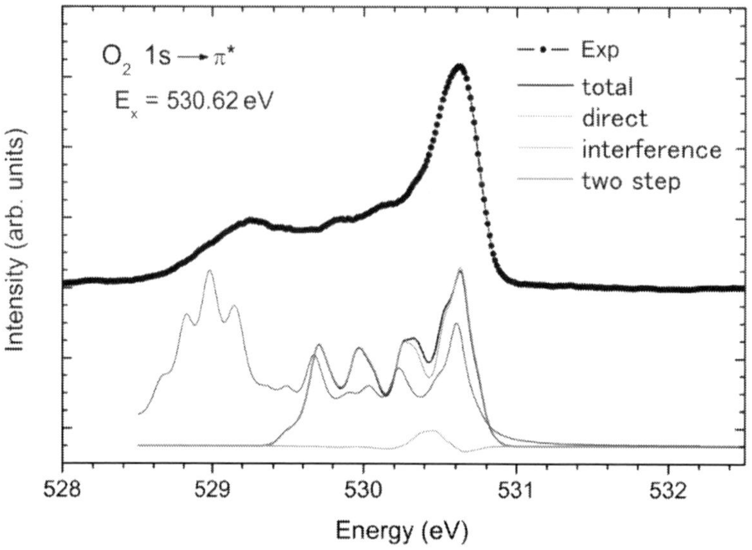

Figure 6. Comparison of the measured and simulated soft x-ray emission (SXE) spectra of O_2 molecule. The solid circles represent the measured spectrum, while the solid lines depict the simulated spectra.

ones, where red curve indicates the total intensity calculated by the one-step formalism, green curve the intensity of interference term in the one-step formalism and blue curve the intensity calculated by the two-step formalism. In the case of N_2 molecule, the resonant excitation to the $v' = 0$ component ($E_x = 400.87$ eV) leads to relatively simple structure. Both of the measured and simulated spectra are showing distinct vibrational structure which can readily be assigned to the $v'' = 0-4$ vibrational structure. In the meantime for the valley excitation at 400.98 eV, i.e., the excitation between two resonances, the simulated spectra indicate that the vibrational components are implying to be split into fine structure, although the measured spectra do not clearly show such multiplication. This is probably due to the lack of resolving power of the SXE spectrometer. For the higher excitation energies ($E_x = 401.32, 401.66$ eV), both of the measured and simulated spectra show much complicated spectra. From the comparison between the one-step and two-step formalisms shown in Fig.5, it seems to be difficult to say which formalism is suitable, but by looking at the high-energy end of the spectra of valley exictation the one-step formalism seems to explain better than the two-step formalism. Although the present simulations could explain a gross feature of the complicated spectra, further refinement, such as account of the relative intensity for vibrational progression, for the simulation may be needed to be done. Detailed measurements and further discussions will be given elsewhere [34].

In the case of O_2 molecule, on the other hand, we did not simulate the SXE spectra in the spectator band because no reliable parameters for the potential energy curve are available for the $3\sigma_g^{-1}\,1\pi_g^3\,{}^3\Pi_g$ final electronic state. Figure 6 indicates the measured and simulated spectra for the excitation energy of 530.62 eV, which corresponds to the resonance excitation to the $v' = 3$ component. As is shown in Fig.1(b), the lifetime width, i.e., 149 meV, of the intermediate core-hole state is larger than the energy spacing of the vibrational progression ($\sim$ 128 meV), the resultant SXE spectra may be affected by the coherent excitation to more than one vibrational level because of stronger overlap between neighboring vibrational levels for the intermediate core-hole states. Thus the $v' = 2$ or 4 components will contribute to and several v'' components will participate in the process, therefore the resultant spectrum will consist of a complex but unresolved structure. It is unfortunate that the resolution of the spectrometer, e.g., 258 meV for the O_2 measurement, was not good enough to resolve such vibrational progression even if we observe the participator band, where the spectroscopic constant ω_e is about 196 meV [26]. According to the fact that the present spectrum has been excited by narrow bandpass radiation ($\sim$ 65 meV), the width of the resultant spectral feature is not limited by the natural width ($\sim$ 149 meV) of the intermediate core-hole state [35]. Apart from the width of the final state, the spectral feature is apparently governed by only the resolving power of the spectrometer. Thus in order to study the resonant SXE processes on O_2 molecule using narrow bandpass radiation, updating of the spectrometer to improve the resolving power is expected. As for the simulation, one-step formalism could not succeed to explain the structure around 529.2 eV but it accounts for the main peak ($\sim$ 530.6 eV) better than two-step formalism. As in the case of the simulation for N_2 molecule, a refinement for the simulation is called for.

4. Conclusions

We have presented the results of measured and simulated vibrationally resolved resonant x-ray emission spectra of N_2 and O_2 molecules in the $1s \rightarrow \pi^*$ resonance region. The simulations based on the lifetime-vibrational interference theory could somewhat explain the gross feature of the fine structure observed in the N_2 spectrum. We find that the resolving power $E/\Delta E \sim 2,000$ for the spectrometer is not sufficient for studying the vibrational structure in the SXE spectrum of O_2 molecule.

Acknowledgments

This experiment was carried out at SPring-8 BL17SU with the approval of the RIKEN SPring-8 Center (Proposal Nos. 20080029 and 20090017). The authors are grateful to Mr. H.Arai for technical assistance.

References

[1] Skytt P, Guo J-H, Wassdahl N, Nordgren J, Luo Y and Ågren H 1995 *Phys. Rev.* A **52** 3572
[2] Skytt P, Glans P, Guo J-H, Gunnelin K, Såthe C, Nordgren J, Gel'mukhanov F Kh, Cesar A and Ågren H 1996 *Phys. Rev. Lett.* **77** 5035
[3] Glans P, Skytt P, Gunnelin K, Guo J-H and Nordgren J 1996 *J. Electron Spectrosc. Relat. Phenom.* **82** 193
[4] Glans P, Gunnelin K, Skytt P, Guo J-H, Wassdahl N, Nordgren J, Ågren H, Gel'mukhanov F K, Warwick T and Rotenberg E 1996 *Phys. Rev. Lett.* **76** 2448
[5] Skytt P, Glans P, Gunnelin K, Guo J-H, Nordgren J, Luo Y and Ågren H 1997 *Phys. Rev.* A **55** 134
[6] Gunnelin K, Glans P, Rubensson J-E, Såthe C, Nordgren J, Li Y, Gel'mukhanov F and Ågren H 1999 *Phys. Rev. Lett.* **83** 1315
[7] Skytt P, Glans P, Gunnelin K, Guo J-H and Nordgren J 1997 *Phys. Rev.* A **55** 146
[8] Tokushima T, Harada Y, Ohashi H, Senba Y and Shin S 2006 *Rev. Sci. Instrum.* **77** 063107
[9] Ohashi H, Senba Y, Kishimoto H, Miura T, Ishiguro E, Takeuchi T, Oura M, Shirasawa K, Tanaka T, Takeuchi M, Takeshita K, Goto S, Takahashi S, Aoyagi H, Sano M, Furukawa Y, Ohata T, Matsushita T, Ishizawa Y, Taniguchi S, Asano Y, Harada Y, Tokushima T, Horiba K, Kitamura H, Ishikawa T and Shin S, 2007 *AIP Conf. Proc.* **879** 523
[10] Senba Y, Ohashi H, Kishimoto H, Miura T, Goto S, Shin S, Shintake T and Ishikawa T 2007 *AIP Conf. Proc.* **879** 718
[11] Tanaka T, Shirasawa K and Kitamura H 2002 *Rev. Sci. Instrum.* **73** 1724
[12] Shirasawa K, Tanaka T, Seike T, Hiraya A and Kitamura H 2004 *AIP Conf. Prof.* **705** 203
[13] Shirasawa K, Hiraya A, Tanaka T and Kitamura H 2004 *Phys. Rev. ST Accel. Beams* **7** 020702
[14] Oura M, Nakamura T, Takeuchi T, Senba Y, Ohashi H, Shirasawa K, Tanaka T, Takeuchi M, Furukawa Y, Hirono T, Ohata T, Kitamura H and Shin S, 2007 *J. Synchrotron Radiat.* **14** 483
[15] Feifel R, Tanaka T, Kitajima M, Tanaka H, De Fanis A, Sankari R, Karlsson L, Sorensen S, Piancastelli M N, Prümper G, Hergenhahn U and Ueda K 2007 *J. Chem. Phys.* **126** 174304
[16] Kato M, Morishita Y, Oura M, Yamaoka H, Tamenori Y, Okada K, Matsudo T, Gejo T, Suzuki I H and Saito N 2007 *J. Electron. Spectrosc. Relat. Phenom.* **160** 39
[17] Tokushima T, Horikawa Y, Harada Y, Takahashi O, Hiraya A and Shin S 2009 *Phys. Chem. Chem. Phys.* **11** 1679
[18] Tokushima T, Harada Y, Takahashi O, Senba Y, Ohashi H, Pettersson L G M, Nilsson A and Shin S 2008 *Chem. Phys. Lett.* **460** 387

[19] Horikawa Y, Tokushima T, Harada Y, Takahashi O, Chainani A, Senba Y, Ohashi H, Hiraya A and Shin S 2009 *Phys. Chem. Chem. Phys.* **11** 8676
[20] Feifel R, Andersson M, Öhrwall G, Sorensen S L, Piancastelli M N, Tchaplyguine M, Björneholm O, Karlsson L and Svensson S 2004 *Chem. Phys. Lett.* **383** 222
[21] Floreano L, Naletto G, Cvetko D, Gotter R, Malvezzi M, Marassi L, Morgante A, Santaniello A, Verdini A, Tommasini F and Tondello G 1999 *Rev. Sci. Instrum.* **70** 3855
[22] Kivimäki A, Maier K, Hergenhahn U, Piancastelli M N, Kempgens B, Rüdel A and Bradshow A M 1998 *Phys. Rev. Lett.* **81** 301
[23] Prince K C, Vondráček M, Karvonen J, Coreno M, Camilloni R, Avaldi L and de Simone M 1999 *J. Electron Spectrosc. Relat. Phenom.* **101-103** 141
[24] Ohashi H, Ishiguro E, Tamenori Y, Okumura H, Hiraya A, Yoshida H, Senba Y, Okada K, Saito N, Suzuki I H, Ueda K, Ibuki T, Nagaoka S, Koyano I and Ishikawa T 2001 *Nucl. Instrum. Methods Phys. Res.* A **467/468** 533
[25] Coreno M, de Simone M, Prince K C, Richter R, Vondráček M, Avaldi L and Camilloni R 1999 *Chem. Phys. Lett.* **306** 269
[26] Farrokhpour H, Alagia M, Coreno M, de Simone M, Prince K C, Richter R, Stranges S and Tabrizchi M 2006 *Phys. Rev.* A **73** 032718
[27] Baev A, Feifel R, Gel'mukhanov F, Ågren H, Piancastelli M N, Bässler M, Miron C, Sorensen S L, Naves de Brito A, Börneholm O, Karlsson L and Svensson S 2003 *Phys. Rev.* A **67** 022713
[28] Huber K P and Herzberg G 1979 *Molecular Spectra and Molecular Structure IV. Constants of Diatomic Molecules* (Van Nostrand Reinhold, New York)
[29] Henke B L, Lee P, Tanaka T J, Shimabukuro R L and Fujikawa B K 1982 *At. Data Nucl. Data Tables* **27** 1
[30] Gel'mukhanov F K, Mazalov L N and Kondratenko A V 1977 *Chem. Phys. Lett.* **46** 133
[31] Murphy R, Lyo I-W and Eberhardt W 1988 *J. Chem. Phys.* **88** 6078
[32] Rubensson J-E, Neeb M, Biermann M, Xu Z and Eberhardt W 1993 *J. Chem. Phys.* **99** 1633
[33] Neeb M, Rubensson J-E, Biermann M, Eberhardt W, Randall K J, Feldhaus J, Kilcoyne A L D, Bradshaw A M, Xu Z, Johnson P D and Ma Y 1993 *Chem. Phys. Lett.* **212** 205
[34] Oura M *et al* (*in preparation*)
[35] Rubensson J-E 2000 *J. Electron Spectrosc. Relat. Phenom.* **110-111** 135

Hydrogen bonding interaction of small acetaldehyde clusters studied with core-electron excitation spectroscopy in the oxygen *K*-edge region

K Tabayashi[1,2,3,4], **M Chohda**[1], **T Yamanaka**[1], **Y Tsutsumi**[2], **O Takahashi**[1,2], **H Yoshida**[1,2,3] **and M Taniguchi**[1,2,3]

[1]*Graduate School of Science, Hiroshima University, Higashi-Hiroshima 739-8526, Japan*
[2]*Faculty of Science, Hiroshima University, 739-8526, Japan*
[3]*Hiroshima Synchrotron Radiation Center, Hiroshima University, Higashi-Hiroshima 739-0046, Japan*

Abstract. In order to examine inner-shell electron excitation spectra of molecular clusters with strong multipole interactions, excitation spectra and time-of-flight (TOF) fragment-mass spectra of small acetaldehyde (AA) clusters have been studied under the beam conditions. The TOF spectra at the oxygen *K*-edge region showed an intense growth of the protonated clusters, M_nH^+ (M=CH_3CHO) in the cluster beams. "cluster-specific" excitation spectra could be generated by monitoring partial-ion-yields of the protonated clusters. The most intense band of $O1s \rightarrow \pi^*_{CO}$ was found to shift to a higher energy by 0.15 eV relative to the monomer band upon clusterization. X-ray absorption spectra (XAS) were also calculated for the representative dimer configurations using a computer modelling program based on the density functional theory. The XAS prediction for the most stable (non-planar) configuration was found to give a close comparison with the cluster-band shift observed. The band shift was interpreted as being due to the HOMO–LUMO interaction within the complex where a contribution of vibrationally blue-shifting hydrogen bonding could be identified.

1. Introduction

Using a synchrotron radiation source, we previously observed inner-shell excitation spectra of formic-acid (HC(O)OH) molecules and small clusters [1] in the beam conditions, and obtained new spectroscopic findings on the local hydrogen bonding (HB) interaction within the molecular clusters as well as fragmentation mechanism upon HB cluster formation. Formic-acid has two oxygen sites available for the strong H-bond formation and forms a stable cyclic-dimer configuration in the gas phase with doubly bridged H-bonds with a binding energy of ~15 kcal/mol [1,2]. The configuration is exclusively the most stable unit-structure among the formic-acid complexes, it may provide the major source of building blocks for the cluster formation. By monitoring partial-ion-yields (PIYs) of the cluster-origin products upon O1*s* excitation, "cluster-specific" excitation spectra with no contribution

[4] To whom any correspondence should be addressed.
E-mail: tabayasi@sci.hiroshima-u.ac.jp

from molecular excitation could be studied. Comparison of the cluster bands with those of molecular formic-acid showed that the first resonance $O1s_{CO} \rightarrow \pi^*_{CO}$ band shifts upwards (to a higher energy) by ~0.31 eV whereas the second $O1s_{OH} \rightarrow \pi^*_{CO}$ band shifts downwards by ~0.82 eV. The band shifts of the core-to-valence transitions in the clusters have been interpreted as being due to the change in core electron binding energy (CEBE) of the oxygen atoms upon HB formation. The $O1s_{CO}/O1s_{OH} \rightarrow \pi^*_{CO}$ band-shifts of the formic-acid clusters have been well supported by a recent density functional theory (DFT) calculation [3].

Acetaldehyde (AA) has a simple structure of C_s symmetry and a high dipole moment, larger than those of typical polar solvents such as water and methanol [4]. Although AA has both hydrogen donor C and lone-pair acceptor O sites within the molecule, strong H-bonds such as the O–H···O are not expected in neutral cluster species. In the case of A–H···O for comparison, strengths of H-bond donor are qualitatively characterized as the differences in electronegativities of the constituent atoms in the associated A–H covalent bond, leading to the classification as F–H > O–H > N–H > C–H [5]. It is thus natural to postulate that the overall structures of small AA clusters are considered as being largely determined by the strong multipole interactions. Governed by such electrostatic interactions, it can be assumed to constitute nearly anti-parallel dimer configurations within the AA clusters.

Here, we have extended inner-shell excitation studies to small acetaldehyde (AA) clusters in the oxygen *K*-edge region, then the effects of such intermolecular interactions on the excitation spectra upon clusterization have been examined. The band-shift of core-to-valence $O1s \rightarrow \pi^*_{CO}$ transition observed in the small AA clusters was also compared with those predicted by the DFT calculation using a computer modeling program [6].

2. Experiments and theoretical calculations

The experiments were carried out in a cluster beam-photoreactive scattering apparatus [7] mounted toa plane-grating monochromator on the soft X-ray beam line BL-6 of the HiSOR facility at Hiroshima University. Two types of sample beams were used; cluster and effusive beams. The cluster beams were generated by supersonic expansion of gaseous mixtures of 1.7–5.0%AA/He via a ϕ 50 μm nozzle, under the stagnation pressures up to 0.3 MPa. The mixtures were prepared by bubbling the He gas through a liquid sample in a cylinder. An effusive beam was prepared by direct introduction of neat AA under the stagnation pressure of P_0 = ~5.0 kPa through a ϕ 200μm nozzle. The stagnation pressures were kept constant by immersing the sample cylinder in a thermostated refrigerant. Either beam was collimated with a ϕ 1.0mm skimmer and allowed to enter the scattering chamber of the apparatus, where it was crossed with a monochromatized soft X-ray beam at the ionization region of a Wiley–MacLaren type [8] TOF spectrometer. Photoelectron-photoion coincidence (PEPICO) signals were recorded using a fast multihit digitizer with a time resolution typically set to 4 ns. The photon energy was calibrated using the $O1s \rightarrow 4p/5s$ Rydberg transition of CO_2 at 540.01 eV [9]. The resolution of the monochromator was $E/\Delta E \geq 2000$ at the energy corresponding to the oxygen *K*-edge. Details of the procedure have been described elsewhere [7].

Acetaldehyde (AA: CH_3CHO) was supplied by Aldrich at a purity of ≥ 99.5% and helium by Nihon-Sanso at ≥ 99.99%. They were used without further purification.

In order to examine the molecular configurations and intermolecular interactions within the AA clusters and support the analysis of core-excitation spectra, ab initio molecular orbital (MO) and DFT calculations were carried out for the AA molecule and clusters. Initial geometries of the system were optimized using the Gaussian 03 program [10] at the MP2/cc-pVTZ level of approximation. Vibrational frequencies were calculated at the same level to confirm their stationary structures and correct the zero-point vibrational energies (ZPVEs). The frequencies were further scaled [11,12] with multiplying by 0.9832 after the MP2 calculations. The basis set superposition error (BSSE) was corrected using the procedures proposed by Boys-Bernardi [13] to estimate the stabilization energies of the clusters.

The calculation of core-excited states was performed with the StoBe-deMon code [6], based on a solution of the Kohn–Sham DFT equations. The theoretical X-ray absorption spectra (XAS) were generated by the transition potential (TP) method combined with a double basis set technique. The orbital energies and oscillator strengths were calculated to obtain the transition energies and absorption intensities in the theoretical XA spectrum. The TP method describes most of the relaxation effects on core-excitation states and provides a single set of orthogonal orbitals for the spectrum calculation. In determining the absolute energy positions of the XA spectrum, ΔKohn–Sham calculations [14] that allow the full relaxation of core-hole ionization states were performed to derive the ionization energy (*IP*). The relativistic correction of 0.33 eV [15,16] was added to the *IP* of the O1*s* electrons. Electron interactions of C and O atoms without any core-electron excitation (core-hole) were described with effective core potentials. The ionized center was described using the IGLO-III basis of Kutzelnigg *et al.* [17], whereas (6311/311/1) and (311/1) basis sets were used for the other heavy atoms and hydrogen atoms. The spectrum was generated by a Gaussian convolution of discrete transition lines by adjusting their broadenings to simulate the experimental spectrum. All DFT calculations were performed using the gradient-corrected exchange and correction functionals developed by Perdew and Wang [18,19]. Further details of the calculations have been given elsewhere [14,20].

3. Results and Discussion

3.1. Core-electron excitation measurements of AA molecule and clusters at the oxygen K-edge

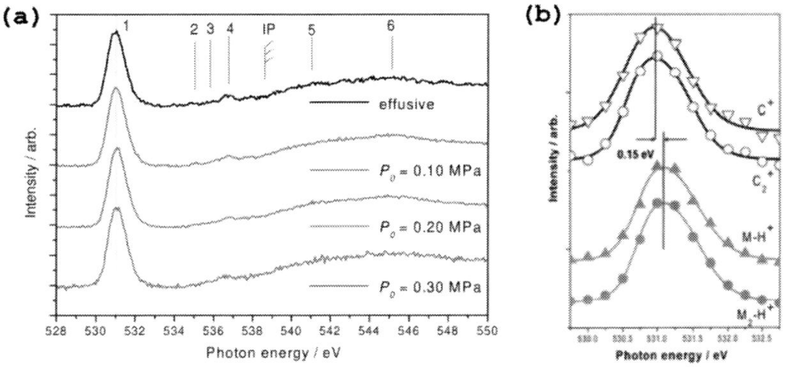

Figure 1. (a) TIY overview spectra of AA molecule (top trace) and clusters (lower three traces) under the stagnation conditions designated in the figure. The band assignment is given in **Table 1**. **(b)** Comparison of the PIY spectra between monomer-origin (C^+, C_2^+) and cluster-origin (MH^+, M_2H^+) products at the O1$s \rightarrow \pi^*_{CO}$ band.

Figure 1(a) shows the typical overview total-ion-yield (TIY) spectra of AA molecule and clusters observed under the beam conditions, recorded in the 528–550 eV excitation energy range that covers oxygen *K*-edge transitions. Publications on the X-ray absorption and excitation spectra of free AA molecule are available in the ISEELS [21] and NEXAFS [22] measurements. The oxygen *K*-edge excitation of AA involves the electron transition from the oxygen –C=O site. Its CEBE has been determined in the X-ray photoelectron spectroscopy [23,24]. Based on the Gaussian deconvolution analysis and previous band assignments, we reassigned the O1*s* excitation bands from the TIY of AA as listed in **Table 1**. In our measurement, closer agreement of the band energies could be obtained with those by Hitchcock and Brion [21]. When comparison of the TIYs in the cluster beams was made with that in the effusive beam, no definite change of the cluster band features for the Rydberg transitions could be admitted under the stagnation pressures up to P_0 = 0.3 MPa, although they are weak and there appears some smoothness in the cluster bands. This is partly because cluster beams are in the small cluster regime where the beams contain significant amounts of uncondensed free molecules as well as the clusters of small sizes.

In the TOF fragment-mass measurements of cluster beams, we observed an intense series of protonated clusters M_nH^+ ($M=CH_3CHO$) under the beam stagnation conditions shown in Figure 1(a). The pressure variation of the product distributions important for the beam characterization has been recently described [25]. It should be added here that, in the effusive beam, trace amounts of MH^+ above the background TOF level could be singly identified as the cluster-origin product. Atomic and smaller fragments such as C^+, O^+, and C_2^+ in the cluster beam show a TOF pattern similar to that of the effusive beam, indicating that they are substantially free AA origin since the beam is again in the small cluster regime. By monitoring the PIY intensities of the M_nH^+ products originating from the clusters, "cluster-specific" excitation spectra could be recorded without any contribution of free molecules. In contrast, the PIYs of atomic and small fragments actually have the same band-peak energy as that observed in TIY of the effusive beam, it is thus reasonable to conclude that they come from free AA molecules. **Figure 1(b)** compares the PIY spectra of the $O1s \rightarrow \pi^*_{CO}$ band for the monomer- and cluster-origin cations simultaneously obtained at the same energy points under the lowest stagnation pressure of $P_0 = 0.1$ MPa. We chose the cluster beam condition of $P_0 = 0.1$ MPa since it provides as small cluster size distribution as possible. Since exclusive intensity of the MH^+ product has been observed over the M_2H^+ and M_3H^+(≈0), it is natural to postulated that the cluster beam at $P_0 = 0.1$ MPa has a smallest cluster size distribution with prominent dimer-contribution. The "cluster-specific" band position was determined with the aid of the deconvolution analysis [26]. A least squares Gausssian-curve fitting of the monomer- and cluster-bands at $P_0 = 0.1$ MPa showed that the cluster $O1s \rightarrow \pi^*_{CO}$ band is broadened slightly and shifted to a higher energy by 0.154±0.010 eV relative to the monomer band. While we tried to examine the upper PIY bands (Rydberg transitions) other than the $O1s \rightarrow \pi^*_{CO}$, reliable data on the relative change of the bands could not be extracted due to low oscillator strengths.

Table 1. Band assignment of acetaldehyde molecule in the oxygen *K*-shell region.

Band	Photon energy *E* / eV			Assignment
No.	Present work	Prince *et al.*[a]	Hitchcook *et al.*[b]	
1	531.02	531.53 (+0.51)	531.14 (+0.12)	π^*_{CO}
2	535.03	535.42 (+0.39)	535.10 (+0.07)	3*sa'*
3	535.81	536.32 (+0.51)	535.84 (+0.03)	3*p*
4	536.77	537.05 (+0.28)	536.68 (-0.09)	4*s*
	538.58[c]	538.58[c]	538.43[d]	*IP*
5	541.34	541.0±0.3	542.0	double exc.
6	547.56	545.2±0.3	544.5	σ^*_{CO}

[a]Prince K C *et al.* (Ref. [18])
[b]Hitchcock A P and Brion C E (Ref. [19])
[c]Jolly W L , Bomben K D and Eyerman C J (Ref. [20])
[d]Jolly W L and Schaaf T F (Ref. [21])

3.2. Computer simulation of XA spectra of AA molecule and dimers derived by DFT calculations

In order to compare the experimental results with the theoretical expectations, and to give an interpretation for the energy shift of the "cluster-specific" $O1s \rightarrow \pi^*$ band from the "monomer" band, some model XA spectra were generated for the representative $(AA)_2$ dimers using the DFT calculation, since the experimental cluster band was recorded under the smallest cluster regime of beam conditions. Examples of the dimer configuration are shown in **Figure 2(b)**; i.e. **dimer 1**: the most stable non-planar (spatial) structure, and **dimer 2**: a planar structure with anti-parallel molecular configuration, optimized at the MP2/cc-pVTZ level of approximation. Some other configurations have also been reported for the dimers using the ab initio MO and DFT calculations [27]. Among these dimers, non-planar spatial structures were found to be more stable than the planar ones by ~0.5–1.0 kcal/mol depending on the level of calculation. Here, electrostatic interactions are firstly important between the AA molecules in determining the stability of the dimer configurations. Under such low symmetric molecules, the system can actually have the closest molecular contact in the non-planar dimer

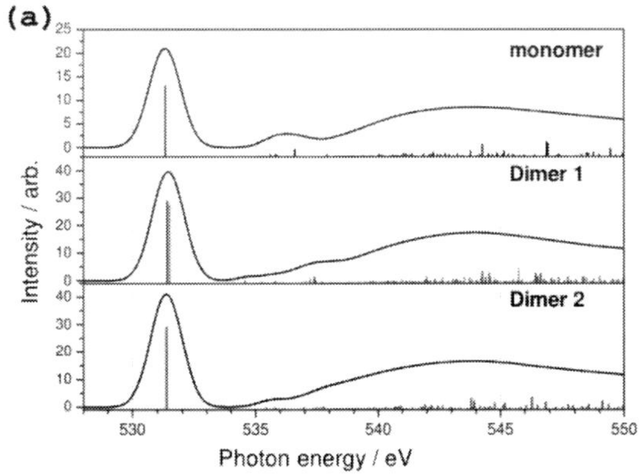

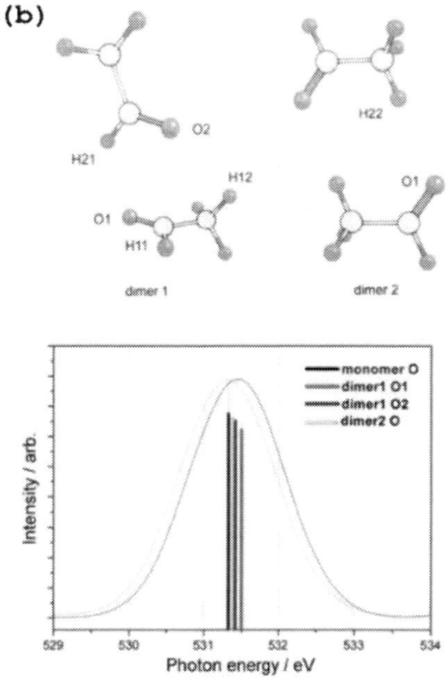

Figure 2. **(a)** Comparison of the calculated XA spectra between the free molecule and representative dimers (**dimer1** and **dimer2**) in the oxygen *K*-edge region. **(b)** Comparison of the calculated O1$s\rightarrow\pi^*_{CO}$ band-spectra between the free molecule and representative dimers.

configurations with high electrostatic stabilization energies. The stability has been attributed to the combination of electrostatic and van der Waals interactions as well as rather weak C–H···O interaction. The binding energy for the most stable **dimer 1** calculated here is $\Delta E_b = -2.70$ kcal/mol and that for the planar **dimer 2** is $\Delta E_b = -1.98$ kcal/mol, including the BSSE and ZPVE corrections.

Table 2. Molecular parameters and core-excitation energies of the representative dimers

System	Interaction	$d_{CH\cdots O}$ / Å	$\theta_{CH\cdots O}$ /deg.	ΔCEBE /eV	Δ(Term v.)[a] /eV	Band-shift[b] /eV
dimer 1	O1–H21	2.386	124.1	– 0.78	– 0.96	0.18
	O2–H11	2.740	75.1	– 0.27	– 0.36	0.09
	O2–H12	2.891	106.6			
dimer 2	O1–H22	2.425	179.2	– 0.46	– 0.44	0.07

[a] Variation of O1$s\rightarrow\pi^*_{CO}$ energy
[b] O1$s\rightarrow\pi^*_{CO}$ band component

Figure 2(a) shows the overview XA spectra calculated for the free molecule and the representative dimers in the O *K*-edge region. Vertical bars represent the oscillator strengths of individual transitions from the constituent O atom in the system. Although the low oscillator strengths of the core-to-Rydberg transitions are reproduced in the calculated spectra, we presently focus on the O1$s\rightarrow\pi^*_{CO}$ transition since reliable comparison with the upper experimental transitions could not be made. **Figure**

2(b) compares the O1s→π^*_{CO} transitions of the representative dimers (**dimer 1** and **dimer 2**) with that of the free molecule. Since O atoms in the non-planar **dimer 1** are not equivalent under their chemical surroundings, their excitation energies have different magnitudes depending on the different influence of the intermolecular (electrostatic etc.) multipoles and local interactions such as the H-bond. The oxygen atoms in the planar **dimer 2**, in turn, are equivalent within the C_{2h} symmetry, so that their excitation energies are actually degenerate. **Table 2** summarizes the molecular parameters (O···H–C) around the O atoms to be core-excited in the dimers, and the excitation energies from the individual O1s orbitals. The energetics are given relative to those of free molecule. In the **dimer 1** configuration, the O1s→π^*_{CO} energies of inequivalent O sites are calculated to be larger than the monomer one (blue-shift) by 0.18 and 0.09 eV, respectively. The largest spectral shift corresponds to the core-excitation of the O atom with the closest O···H contact (**Fig. 2(a)**). These energy shifts are closely consistent with the experimental peak shift (~ 0.15 eV). In contrast, the O1s→π^*_{CO} energy of the planar **dimer 2** shows degenerate transition with a smaller blue-shift by 0.07 eV. The experimental shift cannot be rationalized by the contribution from O1s core-excitation of the planar **dimer 2**. It is thus clear that the experimental results can be strongly explained by the significant contribution from the most stable configuration of the **dimer 1**.

It is notable here that structural optimization analysis of AA clusters larger than the dimers is not available up to this time since AA molecule has such a low symmetric structure and is expected to form a number of configurations as the n-isomers. Recent site-dependent C1s excitation spectra of AA clusters [28], however, shows strongly perturbed C1s_{CO}→Ryd transitions under the relatively higher stagnation pressures at P_0 = 0.2 MPa, indicating the contribution of basic configurations with HCO functional group of the molecule contacting with each other within the clusters. Such configurations of n-isomer should be on a line with that of the **dimer 1**.

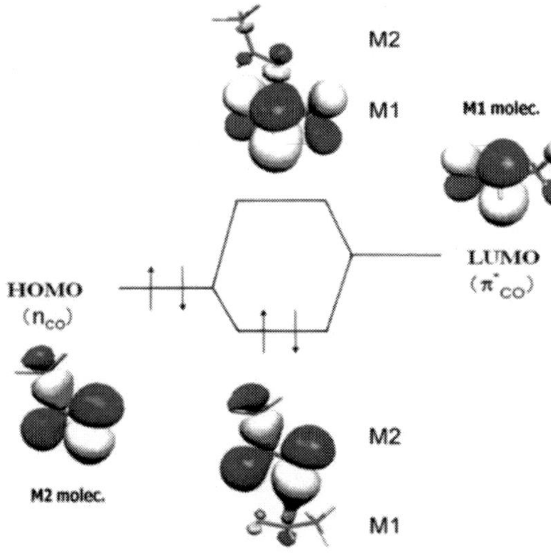

Figure 3. Molecular orbitals showing the HOMO–LUMO interaction within the most stable non-planar **dimer 1**.

From a geometrical view point, the strength of H-bond is formally characterized by H-bond length (or distance) and its directional property [5]. A strong H-bond geometry favours a linear H-bond structure and a O···H–C distance less than the van der Waals contact. The shortest O···H distance (O1–H21) in the **dimer 1** is around ~2.4 Å that shows less than the sum (2.72 Å) of the van der Waals radii of O and H atoms [29]. The C–H···O bond is chemically categorized as a weak H-bond due to the low H-bond donor strength, then its geometries are easily deformed under other strong interactions.

However, the closer O1–H21 contact than the others (O2–H11/O2–H12) may reflect a stronger H-bonding interaction of O1 than the O2, and probably lead to a larger component for the O1 excitation energy. It is notable, in this connection, that the planar **dimer 2** configuration has a favourable geometry despite the weak C–H···O bonds but that sufficient contribution from multipole interaction cannot be gained for the binding energy, due to a longer molecular separation than the **dimer 1**. The definite H-bonding contribution to the $O1s\rightarrow\pi^*_{CO}$ band could not be evaluated since the configuration would be the outcome of the interactions of several types as described above.

Presently, we have observed the spectral band-shift of core-to-valence $O1s\rightarrow\pi^*_{CO}$ transition of small AA clusters in the PIYs of the cluster-origin products. The DFT calculations were applied to the representative dimers to rationalize these experimental results, since the band-shift was determined in the smallest cluster regime of beam conditions. The XAS calculation by DFT showed that the most stable non-planar dimer configuration (**dimer 1**) could well reproduce the blue-shift of the $O1s\rightarrow\pi^*_{CO}$ transition observed. Here, we will focus our attention more details on the $O1s\rightarrow\pi^*_{CO}$ transition in the most stable **dimer 1**.

Table 3. Bond length and vibrational frequency (stretching) calculated by MP2/cc-pVTZ level.

Bond (functional group)	Monomer		The most stable dimer (non-planar)			
			M2 molecule		M1 molecule	
	Length/Å	Freq.[a]/cm^{-1}	Length/Å	Freq.[a]/cm^{-1}	Length/Å	Freq.[a]/cm^{-1}
C–H (CHO)	1.1054	2829	1.1022(–)	2876(+)	1.1028(–)	2855(+)
C=O	1.2130	1706	1.2166(+)	1695(–)	1.2156(+)	1704
C–C	1.4987	1093	1.4969(–)	1096	1.4965(–)	1097

[a] Scaling factor of 0.9575 is employed (Ref. [12]).

In order to consider the origin of the spectral band-shift in the $O1s\rightarrow\pi^*_{CO}$ band, we examined the HOMO–LUMO structures between the constituent molecules M1 and M2 in the most stable **dimer 1**. **Figure 3** shows the HOMO–LUMOs given by StoBe-deMon calculation that expand within the configuration upon dimerzation where the π^*_{CO} orbital of O1 site, corresponding the larger component of the $O1s\rightarrow\pi^*_{CO}$ shift, is designated (the π^*_{CO} orbital of O2 site is not shown). One may easily realize the orbital mixing between the nonbonding n_O and antibonding π^*_{CO} of the present non-planar configuration. The band-shift of the $O1s\rightarrow\pi^*_{CO}$ transition could be thus interpreted as being due to the augmentation of the HOMO–LUMO energy gap as well as the stabilization energy upon dimerization. It is quite important to note that electron transfer to the LUMO, viz. M1 to M2 molecule, is expected to take place under the HOMO–LUMO interaction. Since the LUMO in M2 has a weak σ-bonding nature between the H and C atoms, and another weak π^*_{CO} antibonding nature between the C and O atoms in the CHO functional group, electron transfer to M2 leads to slightly a shorter C–H bond and a longer C=O bond via the HOMO–LUMO interaction. These MO structures of the **dimer 1** imply that there involves the blue-shifting (in vibrational frequency) hydrogen bonding [30] in the interaction. In the ab initio MO calculation, in fact, the present complex (**dimer 1**) formation is found to be accompanied by blue-shifts of the C–H and red-shifts of the C=O stretching frequencies in the HCO functional group in M2/M1. **Table 3** summarizes the bond length and vibrational frequency of the associated bond, where the comparison between the monomer and the most stable **dimer 1** can be made. In the **dimer1** configuration, O1 atom in M1 molecule interacts with C–H (HCO) in M2 molecule and O2 atom in M2 interacts with C-H (CHO) in M1 and so on. The present vibrationally blue-shifting H-bonding causes the contraction of such C-H bond by 0.0016–0.0022 Å and blue-shifts of the C–H stretch frequencies by 26 - 47 cm^{-1} relative to the free molecule. In this respect, the closer

O···H–C contact (in O1–H21 in **Table 2**) is found to lead a stronger blue-shifting H-bond. It is also notable that the configuration of the **dimer 2** has a "proper" (red-shifting) H-bond, although the weak one, where the elongation of the C–H (CH_3) bond by 0.0007 Å and a red-shift of the C-H stretching frequency by 7 cm^{-1} are identified.

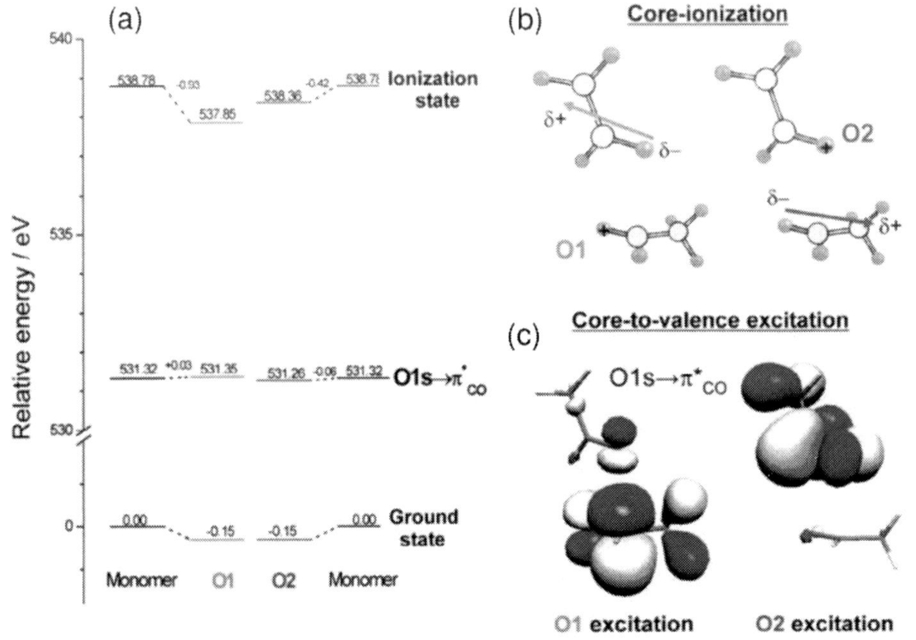

Figure 4. **(a)** Energy level diagram for the core-excited states of individual O atoms in the most stable non-planar **dimer 1**. Energy levels of molecular AA are also shown for comparison. **(b)** A simple charge-dipole approximation for the core-ionization states. **(c)** The $O1s \rightarrow \pi^*_{CO}$ orbitals excited from the individual O1 and O2 sites.

Figure 4(a) compares the energy levels of core-excited states from individual O atoms in the most stable **dimer 1**, with those of AA monomer. Their energetics are given from the DFT calculation. A stable multipole-configuration of the neutral dimer occasionally induces a strong charge-multipole interaction upon core-electron ionization, and stabilizes the core-hole final states. **Figure 4(b)** shows the charge-dipole interaction as a zero-th approximation in the core-ionization states. This simple approximation leads to a larger stabilization energy of O1 ionization than the O2 ionization. The similar trend is also expected for higher Rydberg states. In the core-to-valence transition, however, valence orbitals are not at all diffusing over to the neighbouring molecule, so that interaction between the orbitals is realized when the close contact with the neighbouring molecule can be made. In the present case, it is only when the system forms the non-planar (nearly T-shaped) structure that the n_{CO}–π^*_{CO} (HOMO–LUMO) interaction comes to be important. As described above, the O atoms are not equivalent under their chemical surroundings, and then their excitation energies have different magnitudes depending on the different influence of the intermolecular interactions. With O1 excitation to the $O1s \rightarrow \pi^*_{CO}$ state (**Figure 4 (c)),** the system has the closest O1–H21 contact under the stronger influence of the n_{CO}–π^*_{CO} interaction, leading to the larger $O1s \rightarrow \pi^*_{CO}$ excitation energy. With O2 excitation to the $O1s \rightarrow \pi^*_{CO}$ state, however, less intense mixing of MOs within the dimer can be observed; this is clearly corresponding to the weaker n_{CO}–π^*_{CO} interaction. In the present system where chemical interaction is involved, although the weak one, it is concluded that significant contributions of such orbital interaction to the stabilities of the core-to-valence excited states are made preferentially over the core-hole final state interactions.

4. Conclusions

Core-electron excitation spectra of small acetaldehyde clusters have been studied in the small cluster regime of beam conditions. The "cluster-specific" excitation spectra could be generated by monitoring the PIY spectra of the fragments originating from the clusters. Comparison of the "cluster-specific" band with the "monomer" band revealed that the O1$s \rightarrow \pi^*_{CO}$ band of clusters shifts to a high energy by 0.15 eV. The density functional theory calculation was applied to rationalize the experimental results. As the representative of small acetaldehyde clusters, two types of dimers (the most stable non-planar dimer and anti-parallel aligned planar dimer) were chosen to derive the spectral simulation of the O1$s \rightarrow \pi^*_{CO}$ transitions. The simulation of the most stable dimer with a non-planar geometry could reproduce the experimental band-energy shift upon dimerization, indicative of a significant contribution of the most stable **dimer 1**. The energy shift of the band was interpreted by the augmentation of the HOMO–LUMO energy gap induced in the non-planar complex, where the interaction of vibrationally blue-shifting hydrogen bonding is identified.

Acknowledgments

The present synchrotron radiation experiments were performed at BL-6 at the HiSOR with the approval (#08-A-52) of the Hiroshima Synchrotron Radiation Center (HSRC), Hiroshima University. One of the authors (K.T.) thanks the Research Center for Computational Science at Okazaki for the allocation of SGI Altix4700 computer time.

References

[1] Tabayashi K, Yamamoto K, Takahashi O, Tamenori Y, Harries J R, Gejo T, Iseda M, Tamura T, Honma K, Suzuki I H, Nagaoka S and Ibuki T 2006 *J. Chem. Phys.* **125** 194307
[2] Curtiss L A and Blander M 1988 *Chem. Rev.* **88** 827
[3] Takahashi O, Yamanouchi S, Yamamoto K and Tabayashi K 2006 *Chem. Phys. Lett.* **419** 501
[4] Weast R C *ed.* 2000 *Handbook of Chemistry and Physics* (CRC Press, Boca Raton)
[5] Jeffrey G A 1997 *An Introduction to Hydrogen Bonding* (Oxford University Press, Oxford)
[6] Hermann K, Pettersson L G M and Casida M E 2004 *STOBE-DEMON* Version 2.0, Stobe Software
[7] Tabayashi K, Maruyama T, Tanaka K, Namatame H and Taniguchi M 2007 *AIP Conference Proceedings* **879** 1788
[8] Wiley W C and McLaren I H1955 *Rev. Sci. Instr.* **26** 1150
[9] Prince K C, Avaldi L, Coreno M, Camilloni R and de Simone M 1999 *J. Phys.* B **32** 2551
[10] Frisch M J, Trucks G W, Schlegel H B *et al.* 2004 *GAUSSIAN 03*, Rev. C. 02, Gaussian, Inc., Wallingford, CT
[11] Foresman J B and Frish A 1993 *Exploring Chemistry with Electronic Structure Methods*, 2nd edn., Gaussian Inc.
[12] Sinha P, Boesch S E, Gu C, Wheeler R A and Wilson A K 2004 *J. Phys. Chem.* A **108** 9213
[13] Boys S F and Bernardi F 1970 *Mol. Phys.* **19** 533
[14] Triguero L, Pettersson L G M and Ågren H 1998 *Phys. Rev.* B **58** 8097
[15] Triguero L, Plashkevych O, Pettersson L G M and Ågren H 1999 *J. Electron Spectrosc. Relat. Phenom.* **104** 195
[16] Takahashi O and Pettersson L G M 2004 *J. Chem. Phys.* **121** 10339
[17] Kutzelnigg W, Fleischer U and Schindler M 1990 *NMR-Basic Principles and Progress*, Springer-Verlag, Heidelberg
[18] Perdew J P and Wang Y 1986 *Phys. Rev.* B**33** 8800
[19] Perdew J P and Wang Y 1992 *Phys. Rev.* B**45** 13244

[20] Kolczewski C, Püttener R, Plashkexych O, Ågren H, Staemmler V, Martins M, Snell G, Schlachter A S, Sant'Anna M, Kaindl G and Pettersson L G M 2001 *J. Chem. Phys.* **115** 6426
[21] Hitchcock A P and Brion C E 1980 *J. Electron Spectrosc. Relat. Phenom.* **19** 231
[22] Prince K C, Richter R, de Simone M, Alagia M and Coreno M 2003 *J. Phys. Chem.* A **107** 1955
[23] Jolly W L , Bomben K D and Eyerman C J 1984 *At. Data Nucl. Tables* **31** 433
[24] Jolly W L and Schaaf T F 1976 *J. Am. Chem. Soc.* **98** 3178
[25] Tabayashi K, Chohda M, Yamanaka T, Takahashi O and Yoshida H 2009 *Nucl. Instr. and Meth.* A, in press
[26] Chohda M 2008 *Graduate thesis,* Hiroshima University
[27] Hermida-Ramon J M and Rios M A 1998 *Chem. Phys. Lett.* **290** 431
[28] Tabayashi K *et al.* to be published
[29] Bondi A 1964 *J. Phys. Chem.* **68** 441
[30] Hobza P and Havlas Z 2000 *Chem. Rev.* **100** 4253

International Workshop on Electronic Spectroscopy for Gas-phase Molecules and Solid Surfaces IOP Publishing
Journal of Physics: Conference Series **235** (2010) 012018 doi:10.1088/1742-6596/235/1/012018

X-ray absorption spectra of SiF_4 and $Si(CH_3)_4$ in the Si *K*-shell excitation region

Osamu Takahashi[1], Isao H Suzuki[2], Yutaro Kono[3], Akihiro Ikeda[3], Takao Ouchi[4], Kiyoshi Ueda[4], Itaru Higuchi[5], Yusuke Tamenori[5], and Shin-ichi Nagaoka[3]

[1] Department of Chemistry, Hiroshima University, Higashi-Hiroshima 739-8526, Japan

[2] Institute of Materials Structure Science, High Energy Accelerator Research Organization (KEK), 1-1 Oho, Tsukuba 305-0801, Japan

[3] Department of Chemistry, Faculty of Science, Ehime University, Matsuyama 790-8577, Japan

[4] Institute of Multidisciplinary Research for Advanced Materials, Tohoku University, Sendai 980-8577, Japan

[5] Japan Synchrotron Radiation Research Institute/SPring-8, 1-1-1 Kouto, Sayo-cho, Sayo-gun 679-5198, Japan

e-mail : shu@hiroshima-u.ac.jp

Abstract. X-ray photo-absorption (XA) spectra of SiF_4 and $Si(CH_3)_4$ have been measured in the Si *K*-shell transition region using monochromatized undulator radiation and theoretical calculations have been performed within the framework of density functional theory. For SiF_4 main peak structures of the XA spectrum have been reproduced well with the calculation. Further we theoretically simulated a small peak of Si:$1s \rightarrow 6a_1$ excitation, by sampling distorted geometries using classical trajectory calculations at the ground state, which is a forbidden transition under the dipole selection rule. For $Si(CH_3)_4$ only one peak was observed and reproduced well with the calculation, which was ascribed to the characteristic of broad Rydberg orbitals coupled with valence orbitals.

1. Introduction

Electronic structure and reactivity of compounds including silicon atoms at the molecular level are of great interests due to its importance in the semiconductor industry. During recent development of synchrotron radiation facility, core-level electron spectroscopy has been extensively studied, and understanding on electronic structures and chemical reactions at the atomic level has been advanced. As one of typical examples, site-specific chemical reactions of $SiF_3C_2H_4Si(CH_3)_3$ (FSMSE) in the gas phase have been observed near the Si *L*-edge [1] and especially site-selectivity in an yield of a certain fragment ion was almost 100 %. Furthermore, similar experiments were investigated near the Si *K*-edge [2]. The fragmentation caused through the Si:1*s* ionization is more violent than that caused through the Si:2*p* ionization. Using a variety of techniques such as multi-coinsidence spectroscopies, investigations of FSMSE are continuing for deep understanding on the reaction mechanism of ion fragmentation following core-excitation [3, 4]. Although many spectroscopic studies of molecules including silicon atoms have been performed near the Si *L*-edge, studies near the Si *K*-edge were limited so far [3, 5, 6]. Investigations using similar techniques for related molecules are useful in order to understand the reactivity and the electronic structure of FSMSE efficiently. In the present study, X-ray absorption (XA) spectra of SiF_4 and $Si(CH_3)_4$ as model molecules of FSMSE are investigated experimentally and theoretically, because these molecules correspond to terminal chemical constituents of FSMSE.

SiF_4 and $Si(CH_3)_4$ are simple compounds including a silicon atom with T_d symmetry, and chemical environment around an Si atom is different: An Si atom in SiF_4 is surrounded by four electron-withdrawing groups, on the other hand, that in $Si(CH_3)_4$ is surrounded by four electron-repelling groups. Various spectroscopic data of these simple molecules have been accumulated experimentally [5, 7-10] and theoretically [5, 11, 12]. XA spectra of SiF_4 and $Si(CH_3)_4$ near the Si *K*-edge were reported by several authors [5, 6] and in particular Bodeur *et al.* performed theoretical calculations using *ab initio* configuration interaction (CI) method in the same paper [5]. They discussed valence and Rydberg mixing for double peaks observed in the XA spectra of SiF_4, which were not observed in that of $Si(CH_3)_4$. However, the first small peak observed in that of SiF_4, which is assigned as Si:1*s* → $6a_1$ transition, was not given a sufficient explanation. We are attempting to describe this transition theoretically by taking into consideration geometries distorted from the equilibrium geometry.

A treatment based on the density functional theory (DFT) is now available to inner-shell excitations for analyses of XA spectra and is widely applied to relatively small molecules in the gas phase [13, 14], molecular clusters [15], polymers [16], surface-adsorbed molecules [17], and liquid phase [18, 19]. In the present study, XA spectra of SiF_4 and $Si(CH_3)_4$ near the Si *K*-edge have been measured, and theoretical XA spectra have been produced within the framework of DFT.

2. Methods

2.1. Experimental setup

The experiments were performed on the c branch of the soft X-ray photochemistry beamline BL27SU at the SPring-8 facility [20-22]. The beamline provides linearly polarized monochromatic light ranging from 0.2 to 2.8 keV. The photon energy resolving power during the measurements was set at 3500. The energy calibration of the monochromator was performed with respect to the binding energies of the Ne:1*s* photoelectrons [23]. The intensity of the monochromatized incident photon-beam during the measurement was monitored by collecting the drain current of the post-focusing mirror on the beamline. The experimental XA spectra were measured using a total-ion-yield (TIY) method. The monochromatized photon beam in the Si *K*-edge region was crossed with an effusive beam of sample gases, SiF_4 and $Si(CH_3)_4$, and produced ions were extracted downward with an applied electric field. Ion yields were detected without mass selection using a simple combination of a set of electrodes and micro-channel plates of tandem configuration.

2.2. Theoretical Methods

Firstly, computational methods for the equilibrium geometry are described. Geometry optimizations of SiF_4 and $Si(CH_3)_4$ were carried out using the GAUSSIAN 03 program [24] at the MP2/cc-pVTZ level of approximation. Vibrational frequencies were calculated using the analytical second derivatives at the same level of the geometry optimization.

The XA spectra calculations were performed using the StoBe-deMon program [25]; the detailed computational method was described in previous reports [13, 14, 16]. In brief, the theoretical XA spectra were generated by the transition potential (DFT-TP) method [13]. The gradient-corrected exchange (PD86) [26] and correlation functional (PD91) [27] established by Perdew and Wang were applied. The orbitals for the molecules were determined with a half-occupied core orbital at the ionization site using the IGLO-III basis of Kutzelnigg *et al.* [28] for Si atoms, the (311/211/1) basis set for C atoms, and the (311/1) basis set for H atoms. The augment basis sets on Si atoms were used for Rydberg orbitals: ζ_s=0.017 and 0.0056667, ζ_p=0.014 and 0.0046667, ζ_d=0.015 and 0.003. The auxiliary basis sets were the (5,4;5,4) for Si atoms, the (5,2;5,2) for C and F atoms, and the (3,1;3,1) for H atoms, where the nomenclature ($N_c(s)$, $N_c(spd)$; $N_{xc}(s)$, $N_{xc}(spd)$) indicates the number of *s* and *spd*-type functions to fit the Coulomb and exchange correlation potentials, respectively. The orbitals for the excited electron were then obtained by diagonalizing the Kohn–Sham matrix built from the density with the basis set extended by a large set of diffuse basis functions centered on the excited atom. The obtained orbital energies and computed transition moments provided the excitation energy and associated intensities in the theoretical spectrum. The DFT-TP calculation of the spectrum corresponded to the static exchange approach, and the relaxation effect of the created ion core was neglected by adding the excited electron. This effect is the largest for valence-like excitations, and these states were computed with fully-relaxed Kohn–Sham calculations [14]. For a more accurate estimate of the absolute excitation energy, relativistic and functional corrections were added to the excitation energy [29]. The relativistic corrections added to the ionization potential were 2.0 eV for the Si *K*-edge, whose value was estimated by extrapolating the linear relation between the nuclear charge and the relativistic correction of the second row elements [29]. Functional corrections of 3.6 eV for SiF_4 and 3.8 eV for $Si(CH_3)_4$ were added for the whole spectra. These values were estimated from the difference between the computational and experimental Si:1*s* ionization energies. Finally, the spectra were generated by a Gaussian convolution of the discrete lines by varying the broadenings. The computed spectra of SiF_4 were broadened using Gaussian functions with a constant full width at half maximum (FWHM) of 1.0 eV below 1846 eV, then linearly increasing up to 7.0 eV at 1852 eV. The computed spectra of $Si(CH_3)_4$ were broadened using Gaussian functions with a constant FWHM of 1.0 eV below 1840 eV, then linearly increasing up to 7.0 eV at 1846 eV.

As described below, a symmetry-forbidden transition Si:1*s* → 6a_1 of SiF_4 was observed as the lowest resonant excitation peak in the TIY spectrum. To simulate this transition theoretically, distorted geometries from the equilibrium geometry at the ground state are needed, which are considered to originate from the vibrational motions. To sample them systematically, classical trajectory calculations at the ground state were performed with zero-point vibrational energies without any symmetry constraints. Time step set to 0.5 fs and the calculation was propagated by 10,000 times. Two hundred points in the obtained snapshots were sampled randomly and XA spectra calculations were performed for these selected geometries. Final XA spectra were obtained by a summation of these spectra weighted by the distribution on the ground state because sampled geometries were generally deviated from the equilibrium geometry. The quantum mechanical distribution for the *i*-th vibrational mode $w(i)$ can be calculated using deviations from the equilibrium geometry and the vibrational frequencies,

$$w(i) = \exp(-\sqrt{\omega_i}\,\Delta l_i^2)\,, \qquad (1)$$

where ω_i is the *i*-th vibrational frequency and Δl_i is deviation along the normal mode of the *i*-th vibrational frequency. Note that equation (1) corresponds to the square of the wavefunction of the *i*-th vibrational mode with the vibrational quantum number of zero. Total weight W is defined by the products of $w(i)$,

$$W = \prod_{i=1}^{m} w(i), \qquad (2)$$

where m is the number of vibrational modes. For SiF_4, all vibrational modes contribute in equation (2). For $Si(CH_3)_4$, a few $w(i)$'s become almost zero due to large deviations from the equilibrium geometry for some degrees of freedom. To avoid that the total weight W becomes extremely small, $w(i)$'s larger than a threshold value of 1.0×10^{-3} were only used in this analysis.

3. Results and discussion

SiF_4 and $Si(CH_3)_4$ are of T_d symmetry in the ground state, and electron configuration and unoccupied valence orbitals may be written as follows [8, 10].

SiF_4

$$\underbrace{(1a_1)^2}_{\text{Si }1s}\underbrace{(1t_2)^6(2a_1)^2}_{\text{F }1s}\underbrace{(3a_1)^2}_{\text{Si }2s}\underbrace{(2t_2)^6}_{\text{Si }2p}\underbrace{(4a_1)^2(3t_2)^6}_{\text{inner valence}}\underbrace{(5a_1)^2(4t_2)^6(1e)^4(1t_1)^6(5t_2)^6}_{\text{outer valence}}\underbrace{(6a_1)^0(6t_2)^0}_{\text{unoccupied}}$$

$Si(CH_3)_4$

$$\underbrace{(1a_1)^2}_{\text{Si }1s}\underbrace{(2a_1)^2(1t_2)^6}_{\text{C }1s}\underbrace{(3a_1)^2}_{\text{Si }2s}\underbrace{(2t_2)^6}_{\text{Si }2p}\underbrace{(4a_1)^2(3t_2)^6}_{\text{inner valence}}\underbrace{(5a_1)^2(4t_2)^6(1e)^4(1t_1)^6(5t_2)^6}_{\text{outer valence}}\underbrace{(6a_1)^0(6t_2)^0}_{\text{unoccupied}}$$

TIY spectrum usually approximates a photoabsorption spectrum in the inner-shell excitation of molecules because this excitation generally produces ions with some masses and charges. Figure 1 shows experimental TIY spectra and calculated XA spectra of SiF_4 and $Si(CH_3)_4$ near the Si K-edge. Contrary to the case of the L-shell excitation spectra [8], structures of XA spectra near the Si K-edge become simple due to the absence of the spin-orbit splitting.

XA spectra of SiF_4 and $Si(CH_3)_4$ near the Si K-edge were observed by several research groups [5, 6], and we follow the peak assignments proposed by Bodeur *et al.* and summarized in Table 1. Our experimental peak positions are consistent with Bodeur *et al.* within their experimental error. They discussed assignments around main peaks below the ionization threshold and broad continuum bands

Table 1. Peak positions E (in eV), term values T (in eV), and assignments of XA spectra of SiF_4 and $Si(CH_3)_4$.

SiF_4	Reference 8			Exp.[a]		Calc.[a]		
Peak No.	E_{exp}	T_{exp}	Assignment	E_{exp}	T_{exp}	E_{calc}	T_{calc}	Assignment
1	1846.5	6.0	a_1	1846.1	6.2	1845.6	6.7	$6a_1$
2	1849.0	3.5	t_2	1848.5	3.8	1848.2	4.1	$6t_2$
3	1850.2	2.3	4p	1850.3	2.0	1850.2	2.1	$7t_2$, 4p
IP	1852.5			1852.3[b]		1852.3		
$Si(CH_3)_4$								
Peak No.	E_{exp}	T_{exp}		E_{exp}	T_{exp}	E_{calc}	T_{calc}	Assignment
1	1843.6	2.5	t_2	1843.6	2.7	1843.6	2.7	$6t_2$
IP	1846.1			1846.3[b]		1846.3		

[a] This work. [b] Nagaoka *et al.* (to be published)

above the threshold. However, the first small peak observed in that of SiF_4, which is assigned as Si:1*s* → 6a_1 transition, was not given a sufficient explanation. Our experimental spectral shapes of both molecules are similar to the previous ones due to wide natural broadening even if the high-resolution measurements were performed. For SiF_4, three peaks were observed below the ionization threshold, and theoretical spectrum shown in figure 1(b) is consistent with this result. In the same figure, a line spectrum corresponding to oscillator strength is shown. Note that the origin of a vertical axis of the band spectra is shifted to show an existence of the a_1 symmetry transition clearly. The lowest excited state has no intensity due to the forbidden transition under the dipole selection rule, but this transition seems to occur to some extent (see below). The second peak having the highest intensity can be assigned to Si:1*s* → 6t_2 transition. Although the third peak should be assigned to Si:1*s* → 7t_2 state, Bodeur *et al.* proposed this peak to be assigned as Si:1*s* → 4*p* Rydberg state.

Let us focus ourselves to the lowest excitation here. Under T_d symmetry the transition from the 1a_1 orbital to the lowest unoccupied 6a_1 orbital should not appear in the XA spectra due to the forbidden transition under the dipole selection rule. However, transition to the 6a_1 orbital in SiF_4 F:1*s* excitation [9, 30] and that to the 4a_1 orbital in CH_4 C:1*s* excitation [31] were observed, which originate from symmetry breaking through vibrational motions. The symmetry breaking for either a core orbital or an excited orbital is necessary for activation of this a_1 into a_1 transition. We have attempted to reproduce this transition as follows. Distorted geometries from the equilibrium through classical trajectory calculations were sampled randomly. An unoccupied orbital which was 6a_1 orbital at the equilibrium does not belong to the a_1 symmetry in instances of distorted geometries. XA spectra calculations were

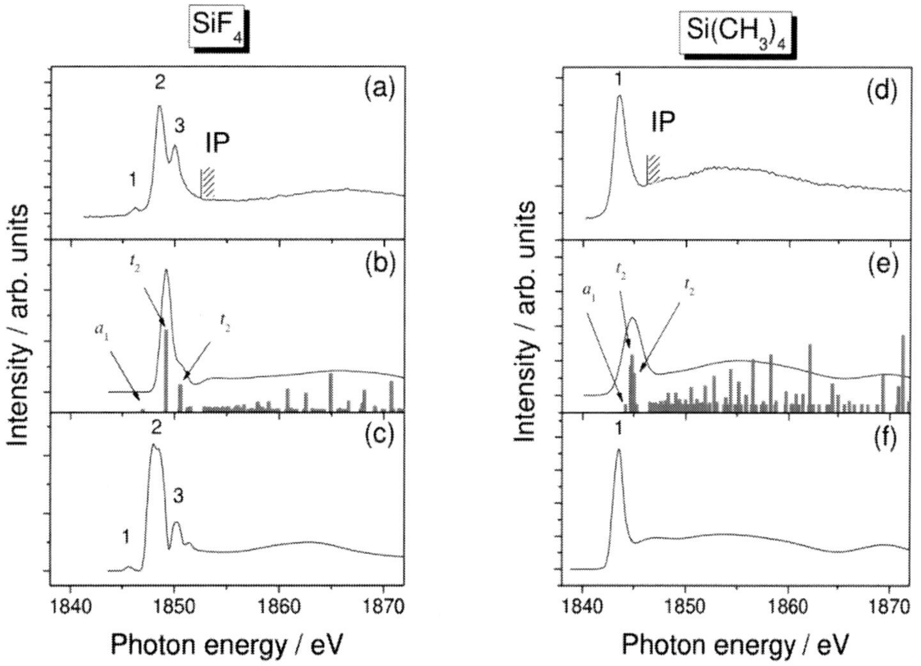

Figure 1. XA spectra of SiF_4 and $Si(CH_3)_4$. (a) and (d) Experimental TIY spectra. Peak numbers and ionization potentials (IP) of an Si:1*s* electron are shown. (b) and (e) Calculated XA spectra for the equilibrium geometry. Some peak assignments are also shown. Line spectrum corresponding to oscillator strength is shown as red lines. (c) and (f) Calculated XA spectra for distorted geometries.

Table 2. Orbital sizes ($<r^2>^{1/2}$) (in Å) for some lower-lying unoccupied orbitals at the core-hole state for SiF_4 and $Si(CH_3)_4$. Those for Si atom are also shown. Corresponding atomic orbitals are shown in parentheses.

	SiF_4	$Si(CH_3)_4$	Si atom
$6a_1$	3.56	7.13	6.32 (4*s*), 8.37 (4*p*), 5.87 (4*d*)
$6t_2$	3.72	7.42	
$7t_2$	9.16	7.34	18.16 (5*s*), 23.57 (5*p*)

performed for more than hundreds geometries and summed spectra with the distribution function *w(i)* were produced. The obtained spectrum was shown in figure 1(c), of which the correspondence with the experimental result is much better than the spectrum for the equilibrium geometry of figure 1(b).

On the other hands, only one peak, which is supposed to originate from many excited states, was observed in the TIY spectrum of $Si(CH_3)_4$. Theoretical spectra of figure 1(e) and (f) have reproduced the experimental spectra well. The lowest $6a_1$ state which was clearly observed as a separate peak for SiF_4, was not observed, because the splitting between a_1 and t_2 orbitals in $Si(CH_3)_4$ is smaller than that of SiF_4; that is, the $6a_1$ an $7t_2$ peaks are buried in the large $6t_2$ peak because of a peak broadening by the geometry relaxation. The present splitting corresponds to 4*s* and 4*p* splitting of Si valence orbitals, and the orbital picture is expected to be asymptotic to that of hydrogen like atom by increasing the principal quantum number. In the case of hydrogen like atom of K_h symmetry, the degeneracy of these two orbitals is due to the characteristic of the coulomb field [32]. Enlargement of the orbital size is often derived from the atomic like character of the orbital, and thus the energy splitting between related orbitals is expected to be small. Table 2 shows spatial sizes of relevant orbitals ($<r^2>^{1/2}$) at the core-hole state for SiF_4 and $Si(CH_3)_4$, and related values for Si atom. $<r^2>^{1/2}$ values for $6a_1$ and $6t_2$ orbitals of SiF_4 are smaller than those of $Si(CH_3)_4$ and Si atom, suggesting that the energy splitting of SiF_4 is larger than that of $Si(CH_3)_4$.

4. Summary

XA spectra of SiF_4 and $Si(CH_3)_4$ in the region of the Si *K*-edge have been studied using synchrotron radiation and theoretical calculations within the framework of the DFT. The calculated spectra are in agreement with the experimental results. In particular the measured Si:1*s* → $6a_1$ transition for SiF_4 have been reproduced well theoretically. As successive studies, we are measuring resonant Auger spectra of SiF_4 and $Si(CH_3)_4$ near the Si *K*-edge for evaluation of excited orbitals. In near future, the electric structures and dissociation processes following the core-excitation of these molecules are being clarified more precisely and being understood more deeply.

Acknowledgements

The authors thank the Information Media Center at Hiroshima University for the use of a grid with high-performance PCs, and Research Center for Computational Science, Okazaki, Japan for the use of a Fujitsu PRIMEQUEST. A part of this study was performed with the approval of the SPring-8 Program Advisory Committee (Proposal No. 2009A-1085).

References

[1] Nagaoka S, Prümper G, Fukuzawa H, Hino M, Takemoto M, Tamenori Y, Harries J, Suzuki I H, Takahashi O, Okada K, Tabayashi K, Liu X J, Lischke T and Ueda K 2007 *Phys. Rev. A* 75 020502R

[2] Nagaoka S, Takemoto M, Prümper G, Fukuzawa H, Tamenori Y, Suzuki I H and Ueda K 2008 *J. Chem. Phys.* 129 204309-7

[3] Nagaoka S, Tamura A, Fujii A, Ohshita J, Okada K, Ibuki T, Suzuki I H, Ohashi H and Tamenori Y 2005 *Int. J. Mass spectrom.* 247 101-5

[4] Suzuki I H, Nitta A, Shimizu A, Tamenori Y, Fukuzawa H, Ueda K and Nagaoka S 2009 *J. Electron. Spectrosc. Relat. Phenom.* 173 18-23
[5] Bodeur S, Millié P and Nenner I 1990 *Phys. Rev. A* 41 252 - 63
[6] Sutherland D G J, Kasrai M, Bancroft G M, Liu Z F and Tan K H 1993 *Phys. Rev. B* 48 14989 - 5001
[7] Urquhart S G, Xiong J Z, Wen A T, Sham T K, Baines K M, de Souza G G B and Hitchcock A P 1994 *Chem. Phys.* 189 757-68
[8] Piancastelli M N, Stolte W C, Guillemin R, Wolska A and Lindle D W 2008 *J. Chem. Phys.* 128 134309
[9] Fujiwara K, Prümper G, De Fanis A, Tamenori Y, Tanaka T, Kitajima M, Tanaka H, Oura M and Ueda K 2005 *Chem. Phys. Lett.* 402 17-20
[10] Sodhi R N S, da Viel S, Brion C E and de Souza G G B 1985 *J. Electron. Spectrosc. Relat. Phenom.* 35 45-64
[11] Hartmann E and Szargan R 1979 *Chem. Phys. Lett.* 68 175-8
[12] Tossell J A and Davenport J W 1984 *J. Chem. Phys.* 80 813-21
[13] Triguero L, Pettersson L G M and Ågren H 1998 *Phys. Rev. B* 58 8097-110
[14] Kolczewski C, Püttner R, Plashkevych O, Ågren H, Staemmler V, Martins M, Snell G, Schlachter A S, Sant'Anna M, Kaindl G and Pettersson L G M 2001 *J. Chem. Phys.* 115 6426-37
[15] Takahashi O, Yamanouchi S, Yamamoto K and Tabayashi K 2006 *Chem. Phys. Lett.* 419 501-5
[16] Takahashi O, Tabayashi K, Wada S, Sumii R, Tanaka K, Odelius M and Pettersson L G M 2006 *J. Chem. Phys.* 124 124901-8
[17] Nyberg M, Odelius M, Nilsson A and Pettersson L G M 2003 *J. Chem. Phys.* 119 12577
[18] Cavalleri M, Ogasawara H, Pettersson L G M and Nilsson A 2002 *Chem. Phys. Lett.* 364 363-70
[19] Wernet P, Nordlund D, Bergmann U, Cavalleri M, Odelius M, Ogasawara H, Naslund L A, Hirsch T K, Ojamae L, Glatzel P, Pettersson L G M and Nilsson A 2004 *Science* 304 995-9
[20] Ishiguro E, Ohashi H, Lu L-j, Watari W, Kamizato M and Ishikawa T 1999 *J. Electron. Spectrosc. Relat. Phenom.* 101-103 979-84
[21] Ohashi H, Ishiguro E, Tamenori Y, Kishimoto H, Tanaka M, Irie M, Tanaka T and Ishikawa T 2001 *Nuclear Instruments and Methods in Physics Research Section A: Accelerators, Spectrometers, Detectors and Associated Equipment* 467-468 529-32
[22] Tamenori Y, Ohashi H, Ishiguro E and Ishikawa T 2002 *Rev. Sci. Instrum.* 73 1588-90
[23] Schmidt V 1997 *Electron Spectrometry of Atoms using Synchrotron Radiation* (Cambridge: Cambridge University Press)
[24] Frisch M J, Trucks G W, Schlegel H B, Scuseria G E, Robb M A, Cheeseman J R, J. A. Montgomery Jr, Vreven T, Kudin K N, Burant J C, Millam J M, Iyengar S S, Tomasi J, Barone V, Mennucci B, Cossi M, Scalmani G, Rega N, Petersson G A, Nakatsuji H, Hada M, Ehara M, Toyota K, Fukuda R, Hasegawa J, Ishida M, Nakajima T, Honda Y, Kitao O, Nakai H, Klene M, Li X, Knox J E, Hratchian H P, Cross J B, Adamo C, Jaramillo J, Gomperts R, Stratmann R E, Yazyev O, Austin A J, Cammi R, Pomelli C, Ochterski J W, Ayala P Y, Morokuma K, Voth G A, Salvador P, Dannenberg J J, Zakrzewski V G, Dapprich S, Daniels A D, Strain M C, Farkas O, Malick D K, Rabuck A D, Raghavachari K, Foresman J B, Ortiz J V, Cui Q, Baboul A G, Clifford S, Cioslowski J, Stefanov B B, Liu G, Liashenko A, Piskorz P, Komaromi I, Martin R L, Fox D J, Keith T, Al-Laham M A, Peng C Y, Nanayakkara A, Challacombe M, Gill P M W, Johnson B, Chen W, Wong M W, Gonzalez C and Pople J A 2004 Gaussian 03, Revision C.02 (Gaussian, Inc., Wallingford CT, 2004).

[25] Hermann K, Pettersson L G M, Casida M E, Daul C, Goursot A, Koester A, Proynov E, St-Amant A, Salahub D R, Carravetta V, Duarte H, Friedrich C, Godbout N, Guan J, Jamorski C, Leboeuf M, Leetmaa M, Nyberg M, Patchkovskii S, Pedocchi L, Sim F, Triguero L and Vela A 2006 StoBe-deMon Version 2.2.
[26] Perdew J P and Yue W 1986 *Phys. Rev. B* 33 8800-2
[27] Perdew J P and Wang Y 1992 *Phys. Rev. B* 45 13244-9
[28] Kutzelnigg W, Fleischer U and Schindler M 1990 *NMR-Basic Principles and Progress* (Heidelberg: Springer-Verlag)
[29] Takahashi O and Pettersson L G M 2004 *J. Chem. Phys.* 121 10339-45
[30] Okada K, Tamenori Y, Koyano I and Ueda K 2002 *Surf. Rev. Lett.* 9 89-92
[31] Ueda K, Okunishi M, Chiba H, Shimizu Y, Ohmori K, Sato Y, Shigemasa E and Kosugi N 1995 *Chem. Phys. Lett.* 236 311-7
[32] Schiff L I *Quantum Mechanics* (Auckland: McGraw-Hill)

Journal of Physics: Conference Series **235** (2010) 012019 doi:10.1088/1742-6596/235/1/012019

Formation of the energetic doubly charged Ne ion by irradiation of large neon clusters using intense EUV-FEL pulses at 52 nm

K Nagaya,[1,2] A Sugishima,[1,2] H Iwayama,[1,2] H Murakami,[1,2] M Yao,[1,2] H Fukuzawa,[2,3] X-J Liu,[2,3] K Motomura,[2,3] K Ueda,[2,3] N Saito,[2,4] A Rudenko,[2,5] M Kurka,[2,6] K-U Kühne,[2,6] J Ullrich,[2,5,6] L Foucar,[2,7] A Czasch,[7] R Dörner,[7] R Feifel,[8] M Nagasono,[2] A Higashiya,[2] T Togashi,[2,9] M Yabashi,[2] T Ishikawa,[2] H Kimura,[9] H Ohashi[9]

[1] Department of Physics, Graduate School of Science, Kyoto University, Kyoto 606-8502 Japan

[2] RIKEN, XFEL Project Head Office, Kouto 1-1-1, Sayo, Hyogo 679-5148, Japan

[3] Institute of Multidisciplinary Research for Advanced Materials, Tohoku University, Sendai 980-8577, Japan

[4] National Metrology Institute of Japan, AIST, Tsukuba 305-8568, Japan

[5] Max Planck Advanced Study Group, CFEL, D-22607, Hamburg, Germany

[6] Max Planck-Institut für Kernphysik, D-69117 Heidelberg, Germany

[7] Institut für Kernphysik, Universität Frankfurt, D-60486 Frankfurt, Germany

[8] Department of Physics and Astronomy, Uppsala University, SE-751 20 Uppsala, Sweden

[9] Japan Synchrotron Radiation Research Institute, Kouto 1-1-1, Sayo, Hyogo 679-5198, Japan

E-mail: nagaya@scphys.kyoto-u.ac.jp

Abstract. The interaction of clusters with intense EUV-FEL pulses was investigated using the SPring-8 Compact SASE Source (SCSS) test facility in Japan. Neon clusters of mean sizes $\langle N \rangle$=1000 and 4000 were irradiated by intense FEL pulses at 52 nm and emitted ions were detected by a momentum imaging spectrometer. The production of energetic doubly charged ions was not found for Ne_{1000}, but it was observed for Ne_{4000} clusters, which suggests that an inhomogeneous charge distribution is generated for the larger clusters.

1. Introduction

The response of matter to intense pulses of laser light in the short wavelength region is an issue of fundamental interest. The recent development of Free Electron Lasers (FEL) based on Self-Amplified Spontaneous-Emission (SASE) in the Extreme UltraViolet (EUV) spectral region below λ=100 nm has enabled us to explore the interaction of intense short wavelength laser pulses with atoms [1–3], molecules [4, 5], and clusters [6–9].

Atomic clusters are ideal objects to study the dynamics driven by the FEL radiation because their size can be controlled from an atom to the bulk-like regime and there is no energy dissipation into surrounding media like in solids. In the present study, we investigated the interaction of neon clusters with intense FEL light. Especially multiple ionization processes were studied at a wavelength of 52 nm by using a dead-time-free multi-particle momentum spectroscopy technique.

2. Experiment

Intense EUV-FEL pulses were provided by the SPring-8 Compact SASE Source (SCSS) test accelerator in Japan [10] with linear polarization in the wavelength region between 51nm and 62nm. A Mg/Si multi-layer mirror fabricated at Lawrence Berkeley National Laboratory (LBNL) was used to focus the FEL pulses into the interaction of a spectrometer set-up. The maximum laser power at the focusing point was estimated to be at most 2×10^{14} W/cm^2 using the estimated (diffraction limited) focal spot size of 3 μm; the experimental conditions considered for this estimation included the laser pulse energy (~30 μJ), the pulse duration (~100 fs), the total reflectivity of three mirrors, the size of skimmers and of a partial light stopper for preventing the incidence of unfocused FEL radiation on the cluster beam [11].

We used a pulsed cluster beam apparatus equipped with a solenoid type pulsed valve (Series 99 valve, Parker Instrumentation Corp.) and a liquid circulating cryostat [12]. The neon cluster beam was produced by supersonic expansion of neat neon gas through a nozzle having an inner diameter of 250 μm. The neon gas was cooled to 85 K using liquid nitrogen and the stagnation pressure was adjusted to control the cluster size. The average number of atoms in the cluster, $\langle N \rangle$, was estimated to be 1000 and 4000 when the stagnation pressure, P_0, was 4.0 atm and 8.6 atm, respectively, by using well-known scaling laws [13, 14].

The momentum imaging spectrometer equipped with a position sensitive delay-line detector [15] was utilized to record the ions emitted from the cluster by FEL irradiation. The produced ions were extracted into the spectrometer using a constant electric field, traveled a second acceleration region and a field free region, and were finally detected by 120 mm diameter microchannel plates (MCPs) with delay-line anode (Roentdek HEX120). The three dimensional momentum vectors for each ion were evaluated from the time-of-flight (TOF) of those ions and their hit positions on the detector. We recorded the signal waveforms from the detector using two 4-channel digitizers (Acqiris DC282), which can substantially reduce the dead times of signals, as compared to a conventional time-to-digital converter system [16].

3. Results and discussion

Figure 1 shows the TOF spectrum of a neon cluster beam ($\langle N \rangle$ = 1000 and 4000) irradiated with intense EUV-FEL pulses at 52 nm with the maximum laser power density. Background signals of residual gases have been subtracted in the spectra of Fig. 1. For a cluster beam with $\langle N \rangle$= 1000, the peaks of singly charged ions (Ne^+ and Ne_2^+) have broad components due to energetic fragment ions in addition to sharp contributions, whereas the peak of Ne^{2+} has only a sharp component. These sharp peaks of Ne^+ and Ne^{2+} stem from the atomic component within the cluster gas jet. The broad peak of Ne_n^+ (n = 1, 2) implies that irradiated Ne clusters become highly charged and subsequently dissociated into many singly charged ions. When P_0 was increased to 8.6 atm, two peaks appear around the atomic Ne^{2+} peak, corresponding to energetic Ne^{2+} ions from the clusters. The striking contrast in peak shape between singly and doubly charged ions suggests the difference to be due to the kinetic energy distribution (KED) among these ions.

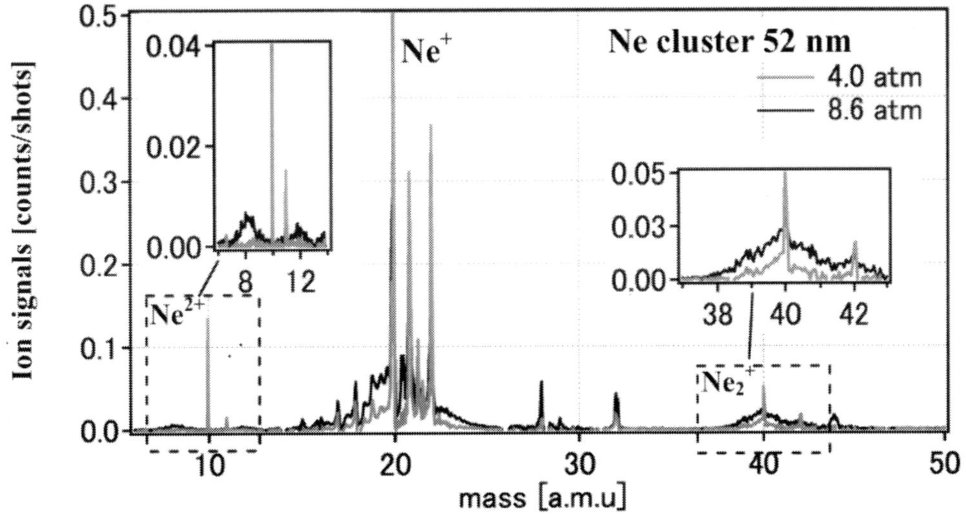

Figure 1. TOF spectrum for Ne clusters irradiated with intense FEL pulses at 52nm. Estimated average cluster size is 1000 and 4000 for the stagnation pressure of 4.0 atm and 8.6 atm, respectively.

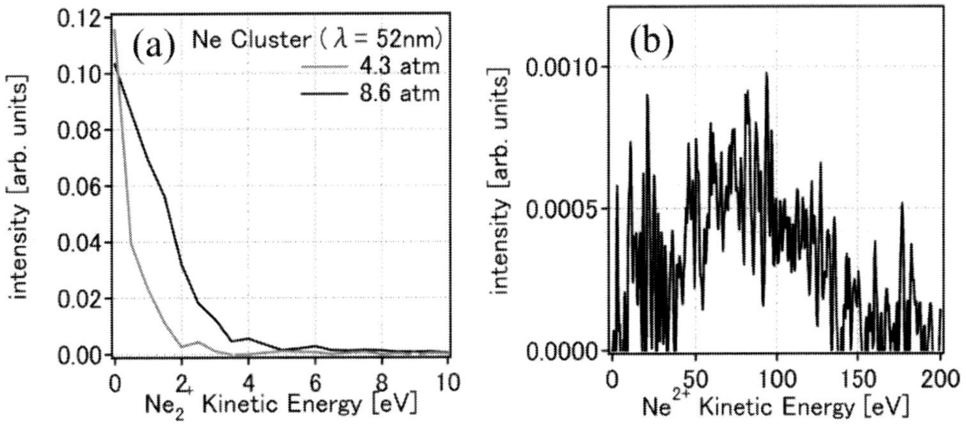

Figure 2. (a). Kinetic energy distribution of the singly charged dimmer, Ne_2^+, from Ne_{1000} (red line) and Ne_{4000} (black line). (b) Kinetic energy distribution of Ne^{2+} emitted from Ne_{4000} clusters irradiated by intense FEL pulses at 52nm.

We evaluated the Kinetic Energy Distribution (KED) of the fragment ions by using the time-of-flight, T, and the hit position, (X,Y), of each ion. To prevent unknown influences of the spectrometer acceptance, we restricted the emission angle of the ions within 25° for the KED analysis [17]. KEDs of Ne_2^+ and Ne^{2+} ions are plotted as a function of kinetic energy in Fig. 2. Here we do not discuss the KED of Ne^+ because we could not evaluate the kinetic energies of individual Ne^+ ions due to overlapping MCP signals from residual gases such as H_2O.

As shown in Fig 2(a), the KEDs of Ne_2^+ monotonically decrease with increasing kinetic energy for both sets of data (4.0 atm and 8.6 atm) displayed, The higher stagnation pressure P_0 makes the average kinetic energy of Ne_2^+ slightly larger. In contrast, the KED of Ne^{2+} has a clear peak around 80 eV, and shows a broad distribution over 100 eV, as shown in Fig. 2(b). Such a difference in the KED between singly and multiply charged ions were also found in the experiments on xenon clusters [12, 18], in which the increase of both, cluster size and laser power density causes the production of the energetic doubly charged ions. The difference in the KED between the singly and highly charged ions could be attributed to different positions of the ions before the dissociation starts by which they are emitted from within a cluster; that is, the highly charged ions are emitted from the cluster surface whereas the singly charged ones are predominantly emitted from the central part of the clusters.

To estimate the relation between the KED and the charge distribution, we have calculated the Coulomb energy stored in the cluster which loses Q electrons under two conditions [18]: One in which the cluster homogeneously contains Q singly charged ions, and another situation in which the cluster contains $(Q-2)$ singly charged ions homogeneously and a doubly charged ion at the cluster surface. By considering the situation where the latter energy is smaller than the former energy, that is the doubly charged ion is stabilized, we obtained the inequality,

$$2(Q-2)e^2/R > 5(I_{12}-I_{01}) - 12\,e^2/R\,.$$

Here, R is the radius of cluster, and I_{01} and I_{12} are the first- and second- ionization potentials of neon atom. The left-hand side of inequality corresponds to the potential energy for a doubly charged ion at the cluster surface, which is converted into its kinetic energy by the dissociation. We have evaluated a characteristic energy for the emergence of doubly charged neon from this inequality as 97 eV, which coincides well with the peak position of the experimental KED of Ne^{2+}. This agreement suggests that an inhomogeneous charge distribution within neon cluster is induced by irradiation with intense FEL pulses.

4. Summary

We investigated the interaction of neon clusters with intense EUV-FEL pulses at 52 nm using the SPring-8 Compact SASE Source (SCSS) test facility in Japan. The Neon clusters were irradiated with intense FEL pulses, and the resulting ions were detected by using a momentum imaging spectrometer. We observed the emission of energetic doubly charged Ne ions when large neon clusters ($\langle N\rangle \sim 4000$) were irradiated by intense FEL pulses at a wavelength of 52 nm. Analyzing the kinetic energy distribution of the emitted ions we conclude that an inhomogeneous charge distribution is created within clusters.

Acknowledgements

We are grateful to the SCSS Test Accelerator Operation Group at RIKEN for their continuous support during the course of our research, to staff of the technical service section in IMRAM, Tohoku University, for their assistance in constructing the apparatus. We are also grateful to A. Belkacem and optics group in LBNL for fabricating the home-made focusing mirror. This study was supported by the X-ray Free Electron Laser Utilization Research Project of the Ministry of Education, Culture, Sports, Science and Technology of Japan (MEXT), a Grant-in-Aids for the Global COE Program "The Next Generation of Physics, Spun from Universality and Emergence" from the MEXT, by Japan Society for the Promotion of Science, and by the MPG Advanced Study Group within CFEL. R.F. acknowledges financial support from the Swedish Research Council (VR) and the Swedish Foundation for International Cooperation in Research and Higher Education (STINT).

References

[1] Sorokin A A *et al* 2007 *Phys. Rev. Lett.* **99**, 213002
[2] Nagasono M *et al* 2007 *Phys. Rev. A* **75**, 051406(R)
[3] Rudenko A *et al* 2008 *Phys. Rev. Lett.* **101** 073003
[4] Föhlisch A *et al* 2007 *Phys. Rev. A* **76** 013411
[5] Sato T *et al* 2008 *Appl. Phys. Lett.* **92** 154103
[6] Wabnitz H *et al* 2002 *Nature (London)* **420** 482
[7] Laarmann T *et al* 2004 *Phys. Rev. Lett.* **92** 143401

[8] Laarmann T *et al* 2005 *Phys. Rev. Lett.* **95** 063402
[9] Bostedt C *et al* 2008 *Phys. Rev. Lett.* **100** 133401
[10] Shintake T *et al* 2008 *Nature Photonics* **2** 555
[11] Fukuzawa H *et al* 2009 *Phys. Rev. A* **79** 031201R
[12] Nagaya K *et al* 2010 *J. Electron Spectr. Related Phenom.* submitted
[13] Hagena O F and Obert W 1972 *J. Chem. Phys.* **56** 1793
[14] Karnbach R *et al* 1993 *Rev. Sci. Instrum.* **64** 2838
[15] Liu X-J *et al* 2009 *Rev. Sci. Instrum.* **80** 053105
[16] Motomura K *et al* 2009 *Nucl. Instr. and Method A* **606** 770
[17] Iwayama H *et al J. Phys.: Conf. Ser. (e2e 2009 proceedings)* accepted
[18] Iwayama H *et al J. Phys. B:At. Mol. Opt. Phys.* submitted

Journal of Physics: Conference Series **235** (2010) 012020 doi:10.1088/1742-6596/235/1/012020

Vibrational Spectra and Geometry Relaxation in Core-Electronic Processes of N_2O and CO_2

Masahiro Ehara[1,2] **and Kiyoshi Ueda**[3]

[1]Institute for Molecular Science, 38 Nishigonaka, Myodaiji, Okazaki 444-8585, Japan
[2]JST, CREST, Sanboncho-5, Chiyoda-ku, Tokyo 102-0075, Japan
[3]Institute of Multidsciplinary Research for Advanced Materisals, Tohoku University, Sendai 980-8577, Japan

E-mail: ehara@ims.ac.jp

Abstract. Vibratinal spectra and geometry relaxation in the core-electron processes of N_2O and CO_2 have been studied by the high-resolution photoelectron spectroscopy and accurate *ab initio* calculations. We calculated reliable two dimensional potential energy surfaces of the core-hole states of N_2O and CO_2 and performed the vibrational analysis based on these surfaces. We also investigated the geometry changes in the π^* and σ^* core-excited states. Theory reproduced the details of the high-resolution spectra providing quantitative assignments and interpretations.

1. Introduction

Vibrational excitation often appears in the photo-ionization and excitation of molecules. This is also the case for core-level photoemission/absorption, even though the core electrons are non-bonding and therefore a significant change in geometry may not be expected. Recent high resolution molecular spectroscopy can observe the fine structure of these vibrational spectra in the core-electronic processes. Theoretical information is valuable to interpret these high-resolution vibrational spectra and to clarify the physics behind the spectra.

Thus, the vibrational spectra and geometry relaxation in the core-hole/core-excited states of various molecules have been investigated by both experiment and theory. For example, for CO_2, the vibrational spectra of the C 1s and O 1s ionized states were measured by Carroll et al. [1] and Kivimäki et al. [2]. Theoretically, Hahne et al. studied the geometry change and vibrational frequency in the O 1s ionized state with the MP2 method [3]. Dobrodey et al. provided the branching ratio in the vibrational levels of O $1s^{-1}$ with the SDCI method [4]. Previously, we also investigated the vibrational structure observed in the core-hole states of N_2O [5] and CO_2 [6]. For the inner-shell excited states of these molecules, several experimental and theoretical works have also been performed [7-14]. In the previous studies, we investigated the vibrational spectra in the O 1s excited states of N_2O and clarified the origin of unusual vibrational progression [9] and its thermal effect [10]. We also studied the geometries in the C 1s and O 1s excited states of CO_2 and in the N 1s excited states of N_2O [11]. However, the σ^* core excited states have not been focused for these molecules.

The SAC-CI method [15,16] has been successfully applied to wide varieties of chemistry and physics. The method has been established through numerous applications to molecular spectroscopy, biological chemistry, and surface chemistry [17]. The SAC–CI general-*R* method [18,19] was developed for investigating complex multi-electron processes like shake-up satellite states and has been successfully applied to those processes [20]. Since the method is designed for the general electronic states, it is also useful for the inner-shell electronic processes. We have investigated various kinds of core-electronic processes using the general-*R* method [6,9-11,21-27]; for example, core-electron binding energies (CEBE), inner-shell shake-up satellite spectra, vibrational spectra and geometry relaxation of satellites, g–u splitting, valence–Rydberg coupling, and its vibration-induced suppression.

In this article, we present the theoretical and experimental studies on the vibrational spectra and geometry relaxation in the core-electronic processes of N_2O and CO_2. We first explain our *ab initio* calculations and vibrational analysis using the grid method. Then, we discuss the vibrational spectra in both the core-hole and core-excited states of N_2O and CO_2. Previously, we performed the vibrational analysis of the core-hole states of these molecules. In this paper, we present reliable two dimensional (2D) potential energy surfaces of these core-hole states. We also present the geometry changes in both π^* and σ^* core excited states of these molecules.

2. Computional details

2.1. *Ab initio* calculations

We adopted two approaches to calculate potential energy surfaces of the core-hole states. One is the SAC-CI method which calculates the core-hole states directly, while the other is the coupled-cluster singles and doubles with noniterative triples CCSD(T) method within the equivalent core approximation (denoted as ECA CCSD(T)) where the core-ionized atom is replaced by (Z=*N*+*1*) atom and the core-hole state is calculated as the ground state. For core-excited states, we directly calculated the excited states by the SAC-CI method.

In the SAC–CI calculations, we adopted the general-*R* method in which the neutral ground-state Hartree–Fock wave function is used for the reference function. The wave function for core ionization, for example, is given by,

$$\Psi^{SAC-CI}_{general-R} = \left(\sum_i R_i + \sum_{ija} R^a_{ij} + \sum_{ijkab} R^{ab}_{ijk} + \sum_{ijklabc} R^{abc}_{ijkl} + \cdots \right) \exp\left(\sum_I S_I \right) \Phi_{HF}, \tag{1}$$

where *R* and *S* represent the *R*- and *S*-operators in the SAC–CI expansion, respectively, and Φ_{HF} is the closed-shell ground-state Hartree–Fock wave function. In this method, the higher *R*-operators such as triples and quadruples are necessary for describing both orbital relaxation and electron correlations of the core-electronic processes. Although the SD-*R* method is accurate for the one-electron processes of the valence excitations and ionizations, higher-order operators are indispensable for the core-electronic processes.

2.2. Vibrational analysis

We have performed the 2D vibrational analysis based on the *ab inito* SAC-CI general-*R* or ECA CCSD(T) potential energy surfaces of the inner-shell electronic states. For calculating the vibrational spectrum, vibrational states were obtained by the grid method, in which Lanczos algorithm was adopted for the diagonalization. Since N_2O and CO_2 have linear structures in the core-hole states, we calculated the symmetric and anti-symmetric stretching modes. Thus, the 2D potential energy surfaces of triatomic molecules were described in the binding coordinates, where the bond distances are r_1 and r_2. In these coordinates the kinetic part of the Hamiltonian of the vibrational motion in molecule A-B-C is given by

$$T = -\frac{\partial^2}{2\mu_{AB}\partial r_1^2} - \frac{\partial^2}{2\mu_{BC}\partial r_2^2} + \frac{\partial^2}{m_B \partial r_1^2 \partial r_2^2} \tag{2}$$

where μ_{AB} is the reduced mass of A and B atoms. The coordinates r_1 and r_2 were represented by the Hermite discrete variable representation (DVR). The potential energy surfaces were fitted by the 2D-Morse expansion,

$$V(r_1,r_2)=\sum_{i,j=0}^{n} B_{ij}(1-e^{-a_1(r_1-r_{1e})})^i(1-e^{-a_2(r_2-r_{2e})})^j \tag{3}$$

where r_{1e} and r_{2e} are equilibrium distances in each state and were determined by the analytical energy gradients of the SAC-CI or ECA CCSD method. Finally, the vibrational spectra were calculated with the Franck-Condon (FC) approximation. The calculations of 2D vibrational wave functions were executed using the MCTDH program system developed by the Heidelberg group [28].

3. Core-hole states of N_2O and CO_2

3.1. N_2O

N_2O is a linear molecule that has two nitrogen atoms, labeled as center (N_c) and terminal (N_t). We have observed vibrational structures both for the N_c and N_t core-level photoelectron spectra and analyzed the vibrational spectra by calculating the 2D potential energy surface in the symmetric and anti-symmetric stretching coordinates [5]. Relative intensities for vibrationally resolved photoelectron spectra were calculated with the FC approximation. Intensities were also examined using the multi-channel Schwinger configuration interaction method (MCSCI) within the adiabatic approximation.

The N_c 1s and N_t 1s photoelectron spectra of N_2O, in the form of $I(0°)+2xI(90°)$, measured at a photon energy $h\nu$=450 eV, are shown in Fig. 1. In N_c 1s photoionization, one vibrational progression, quasi-symmetric stretching mode υ'_1 was enough to fit the spectrum. For N_t 1s spectrum, on the other hand, two vibrational progressions, both quasi-symmetric and quasi-antisymmetric modes were necessary to get reasonable curve fit.

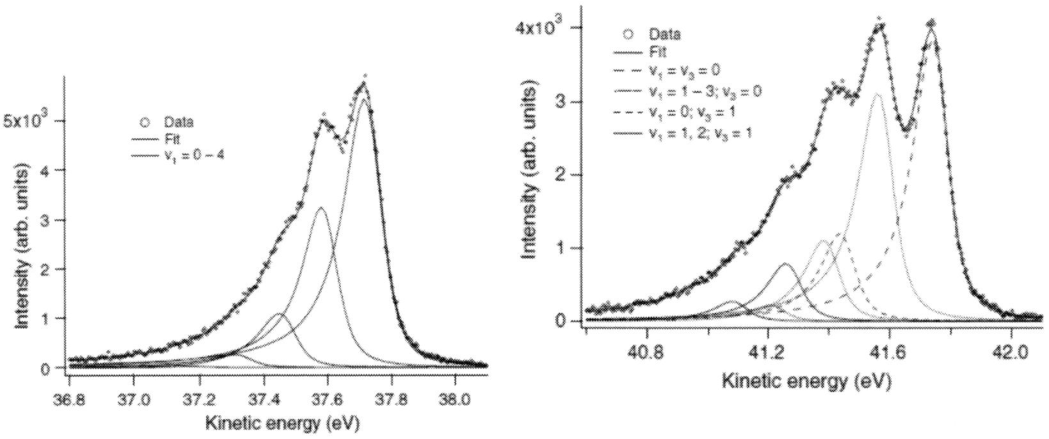

Figure 1. N_c 1s (left) and N_t 1s (right) photoelectron spectra $I(0°)+2xI(90°)$ of N_2O measured at $h\nu$=450eV [5].

The 2D potential energy surfaces of the ground and N 1s ionized states were calculated by the ECA CCSD(T) method. The calculated potential energy surfaces of these states are displayed in Fig. 2 with the equilibrium structures. Geometry change is large in N_t $1s^{-1}$ state than that in N_c $1s^{-1}$ state, which explains the vibrational progressions in these core-hole states shown in Fig. 1. The potential energy surfaces of the N_t $1s^{-1}$ and N_c $1s^{-1}$ states are flat in comparison with that of the ground state.

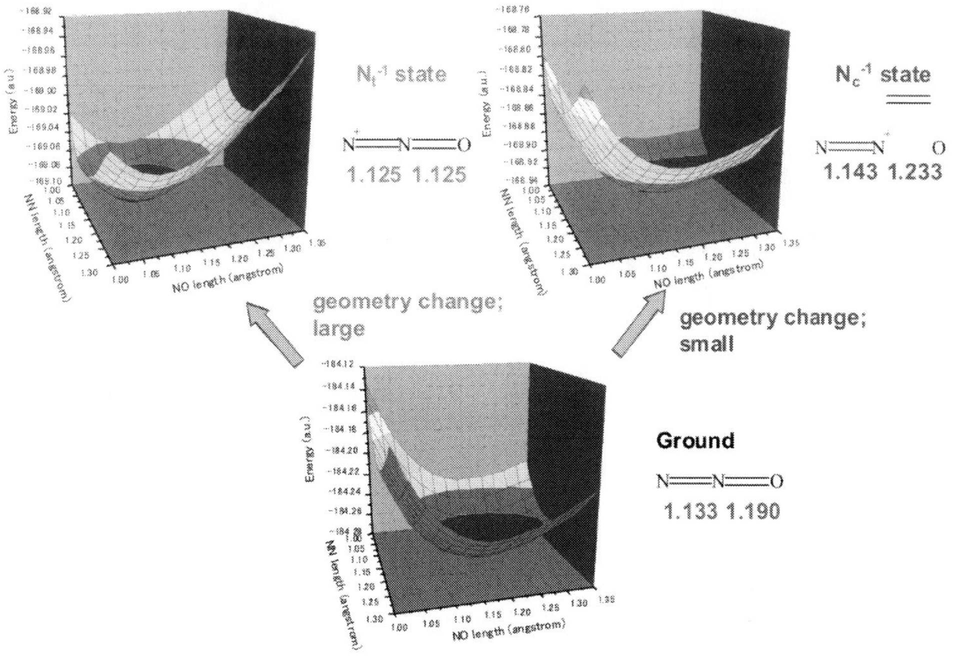

Figure 2. 2D potential energy surfaces of the ground and core-ionized states, N_c $1s^{-1}$ and N_t $1s^{-1}$ states.

The geometrical parameters and the spectroscopic constants of these states are summarized in Table 1 together with the ratios of calculated FC factors and the experimental photoionization ratios. We note that the intensity ratios of the vibrational components are correlated to the ratios of the FC factors when the excitation energy is sufficiently high, that is in sudden limit. We have calculated the FC factors ratios based on the CCSD(T) potential.

The geometry change is sensitive to the level of the theoretical calculations. We present here the results of CCSD(T). According to the CCSD(T) calculation, the equilibrium structures of the ground state and core-ionized states are linear. The optimized bond lengths of the ground state are R_{NN}=1.133 and R_{NO}=1.190 Å, which well agree with the experimental values of R_{NN}=1.127 and R_{NO}=1.185 Å [29]. The CCSD(T) calculations indicate that both NN and NO bonds are elongated relative to the ground state, ΔR_{NN}=+0.010 and ΔR_{NO}= +0.043 Å, in the N_c 1s ionized state. For the N_t 1s ionized state, on the other hand, the CCSD(T) calculations predict both NN and NO bonds shrinks, ΔR_{NN}=−0.008, ΔR_{NO}=−0.065 Å: the change in R_{NO} is much larger than that in R_{NN}.

The vibrational spectrum of the N_c 1s ionized state calculated by CCSD(T) reasonably agrees with the experimental spectrum. The vibrational excitations are dominantly due to υ'_1 mode. The calculated vibrational intensity ratios were 0.47 and 0.11 for (υ'_1,υ'_3)=(10) and (20), respectively, in comparison with the respective experimental values of 0.60 and 0.20. This suggests the underestimate of the geometry change. For the N_t 1s ionized state, on the other hand, the theoretical spectrum by CCSD(T) satisfactorily agrees with the experimental spectrum. The vibrational frequencies are calculated to be 175 and 295 meV for υ'_1 and υ'_3, respectively, in comparison with the experimental values of 178 and 298 meV. The calculated intensity ratios are 0.84, 0.30, 0.28 and 0.19 for (10), (20), (01) and (11), respectively, and are in reasonable agreement with the respective experimental ratios of 0.82, 0.29, 0.31 and 0.21.

Table 1. Vibrational constants, Franck-Condon factor ratios $R(\upsilon'_1, \upsilon'_3)=I(\upsilon'_1, \upsilon'_3)/I(0,0)$ and the geometry change for the N_c1s and N_t1s ionized states of N_2O [5].

	Expt.	ECA CCSD(T)		
			FCF	MCSCI[a]
N_c				
ω_{e1} (meV)	136(2)	125		
ω_{e2} (meV)	-	225		
R(1,0)	0.60(1)		0.467	(0.472)
R(2,0)	0.20(6)		0.110	(0.111)
R(0,1)	-		0.002	(0.004)
R(1,1)	-		0.000	(0.002)
ΔR_{NN} (Å)	-	+0.010		
ΔR_{NO} (Å)	-	+0.043		
N_t				
ω_{e1} (meV)	178(1)	175		
ω_{e2} (meV)	303(2)	295		
R(1,0)	0.82(1)		0.840	(0.819)
R(2,0)	0.29(2)		0.301	(0.287)
R(3,0)	0.05(2)		0.061	(0.057)
R(0,1)	0.31(2)		0.276	(0.289)
R(1,1)	0.21(2)		0.192	(0.194)
R(2,1)	0.07(2)		0.055	(0.054)
ΔR_{NN} (Å)	-	-0.008		
ΔR_{NO} (Å)	-	-0.065		

[a] Photoionization ratios calculated with multi-channel Schwinger configuration interaction.

3.2. CO_2

The vibrational spectra of C 1s and O 1s ionization of CO_2 were also studied. This system is of interest since it shows dynamic core-hole localization in O 1s ionization. Theoretically, vibrational frequency and FC factors/geometry relaxation were studied by MP2 [3] and SDCI [4] in the previous works.

C 1s and O 1s photoelectron spectra of CO_2, measured at a photon energy $h\nu$=500 and 700 eV, respectively, are shown in Fig. 3 together with a least-squares curve fitting decompositions [6]. The vibrational structure in the C 1s spectrum was assigned to the symmetric stretching vibrations (υ'_1, 0, 0), whereas the progression in the O 1s spectrum was assigned to the antisymmetric stretching vibrations (0, 0, υ'_3). Antisymmetric stretching vibration is excited by the vibronic coupling and provides a mechanism of dynamic core-hole localization.

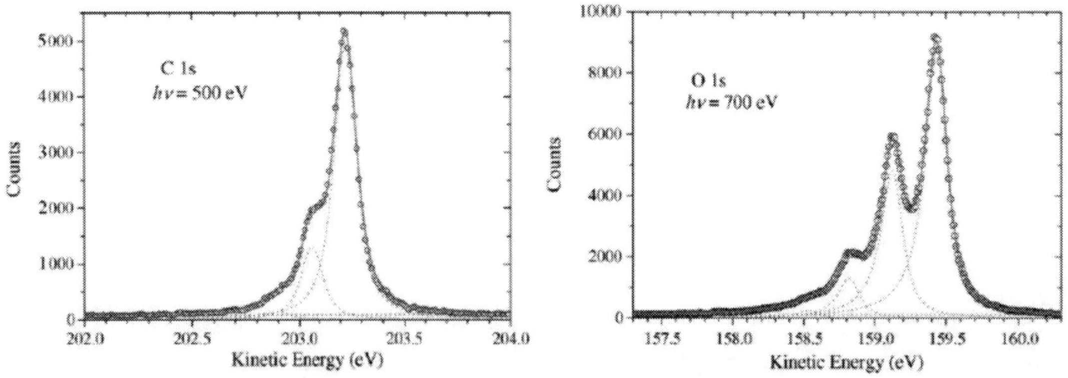

Figure 3. C 1s and O 1s photoelectron spectra of CO_2 measured at 500 and 700 eV, respectively [6].

We calculated the two dimensional potential energy surfaces of the ground, C 1s and O 1s ionized states by using both the ECA CCSD(T) and SAC-CI general-*R* methods. The calculated 2D potential enegy surfaces are shown in Fig. 4. For O $1s^{-1}$, we show one of the two Jahn-Teller local minima and its diabatic surface; the other surface in which 1s electron in the other oxygen is ionized exists. These two surfaces are relevant in the dynamic localization of core hole of O1s ionized states of CO_2. As shown in figure, symmetric mode is excited in C $1s^{-1}$, while anti-symmetric mode is excited in O $1s^{-1}$.

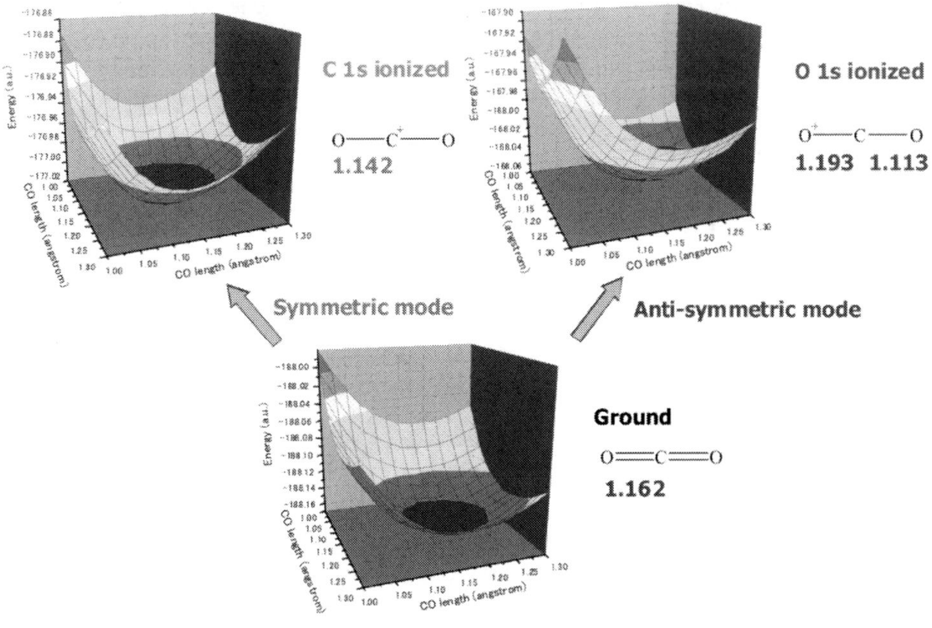

Figure 4. 2D potential energy surfaces of the ground and core-ionized states, C $1s^{-1}$ and O $1s^{-1}$ states.

Table 2. Vibrational constants, Franck-Condon factor ratios $R(\upsilon'_1, \upsilon'_3)=I(\upsilon'_1, \upsilon'_3)/I(0,0)$ and the geometry change for the C 1s and O 1s ionized states of CO_2 [6].

	Expt.	Theoretical	
		SAC-CI	ECA CCSD(T)
C 1s			
ω_{e1} (meV)	164(1)	175	169
ω_{e2} (meV)	0.2(1)	0.2	-
$R(1,0)$	0.262(4)	0.278	1.139
$R(2,0)$	0.028(4)	0.016	0.536
ΔR_{CO} (Å)	-0.0212	-0.020	-0.049
O 1s			
ω_{e1} (meV)	309(1)	330	302
$R(1,0)+R(2,0)$	0.553(6)	0.956	0.534
$R(2,0)+R(2,1)$	0.151(4)	0.349	0.127
$R(3,0)+R(2,2)$	0.028(4)	0.057	0.020
ΔR_{CO+} (Å)	0.0427(3)	0.031	0.040
ΔR_{CO} (Å)	-0.0427(3)	-0.049	-0.043

The calculated spectroscopic constants of the ground state are in good agreement with the well-known experimental values. The calculated spectroscopic constants of the C 1s and O 1s ionized states are summarized in Table 2. We first discuss the SAC-CI general-*R* results. In the C 1s ionized state, the calculated CO bond length decreases by $\Delta R_{CO}=-0.020$ Å from that of the ground state in very good agreement with the experimental value of $\Delta R_{CO}=-0.021$ Å. The calculated frequency $\omega_{e1}=175$ meV

Journal of Physics: Conference Series **235** (2010) 012020 doi:10.1088/1742-6596/235/1/012020

and anharmonicity $\omega_{e1}\chi_{e1}$=0.2 meV are in reasonable agreement with the measurement of ω_{e1}=164 meV and $\omega_{e1}\chi_{e1}$=0.2 meV, respectively. The CO^+ bond length with an O1s ionization was, on the other hand, calculated to be longer by 0.031 Å than that in the ground state, whereas the other CO bond length is shorter by −0.049 Å. The estimated change of the bond length is in fair agreement with the experimental estimate of ΔR_{CO}=±0.043 Å.

The FC factors were also calculated from the vibrational wavefunctions and the calculated FCF ratios are listed in Table 2 together with the experimental values. The SAC-CI values are in good agreement with the experimental values for C 1s while the agreement becomes worse for O 1s. We also performed the ECA CCSD(T) calculation. The FCFs of the O 1s ionized state by the ECA CCSD(T) method agree well with the experimental values, though the agreement is worse for the C 1s ionization. The SAC-CI general-*R* method can be applied to wide varieties of core electronic processes and usually provides reliable results. However, since it uses the ground-state Hartree-Fock for reference, higher-order *R*-operators are necessary for describing orbital relaxation. Therefore, for the system that exhibits large orbital relaxation, such as O 1s ionized CO_2 concerned here, ECA CCSD(T) method can give better result than the SAC-CI general-*R* method.

4. Core-excited states of N_2O and CO_2

The equilibrium structures of the core-excited states of N_2O and CO_2 are characteristic. The core-excited Π state of these molecules is doubly degenerate in the linear geometry and splits into two states along the bending coordinate due to Renner-Teller effect. We studied both Π and Σ states of these molecules.

The N1s-π^* excited states have a bent structure; qualitatively, these states are compared with ONO and NOO molecules in the equivalent core model for N_t and N_c excitations, respectively. The optimized structures of these states are displayed in Fig. 5. For the N_t 1s-π^* excited state, the calculated bond lengths and angle were R_{NN}=1.192 Å, R_{NO}=1.198Å and 133.9°, respectively, in good agreement with the ECA experimental values of ONO, 1.1945Å and 133.9°. The geometry change is more drastic in the N_c excited state. The bent structure was also found in the O 1s-π^* excited state. We also calculated the equilibrium geometry of the N_t 1s-σ* and N_c 1s-σ* excited states, although they are located above the respective core-hole states. The geometries of these states are both linear. In N_t 1s-σ*, the N-O bond length shrink, while in N_c 1s-σ*, the N-O bond length becomes large.

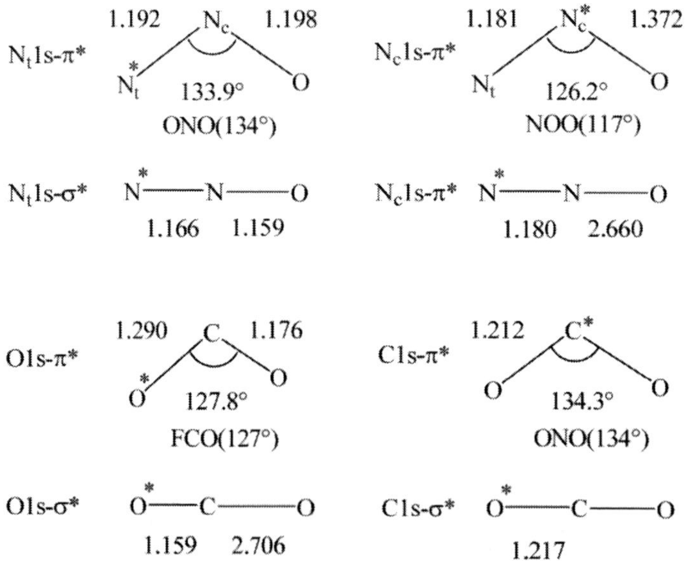

Figure 5. Calculated equilibrium geometries (Å) of the ground and core-excited states of N_2O and CO_2.

Next, we discuss the geometry in the C 1s and O 1s excited states of CO_2. The calculated equilibrium structures in these states are also displayed in Fig. 5. The π^* core-excited states of CO_2 also have bent structures due to the Renner-Teller effect. Previously, the vibrational spectrum of the C 1s-π^* state was observed by XPS and the geometry of this state was experimentally probed [12,13]. In the C 1s-π^* excitation, the bond lengths and angle relax to 1.212Å and 134.3°, respectively. Compared with ONO, the equivalent-core molecule, the bond angle is almost the same. The MRSDCI method also calculated the bond distances and angle in accord with the present values [14]. The stable structure of the O 1s-π^* excited state of CO_2 is also nonlinear, but asymmetric: the calculated bond lengths of CO* and CO were 1.290Å and 1.176Å, respectively, and the bond angle was 127.8°. The σ* core-excited states also have a minima in the linear structure. The bond lengths are elongated in C 1s-σ* state, while in O 1s-σ* state, C-O* bond shrinks and the other C-O bond length becomes large.

5. Summary

We have investigated the vibratinal spectra and geometry relaxation in the core-electron processes of N_2O and CO_2 by the high-resolution photoelectron spectroscopy and accurate *ab initio* calculations. For core-hole states of these molecules, we calculated reliable 2D potential energy surfaces by the ECA CCSD(T) and SAC-CI general-*R* methods. The calculated vibrational spectra with FC analysis reproduced the experimental photoelectron spectra providing the quantitative assignments. The geometry relaxation in the core-excited states of these molecules was also investigated for the transition to both π* and σ* orbitals. The equilibrium geometries in π* core-excited states are bent, while they are linear in σ* core-excited states.

Acknowledgments

The authors sincerely thank Prof. H. Nakatsuji for all the collaboration in the present article. They also thank Dr. T. Tanaka and Prof. H. Tanaka for the high-resolution experiment. This study was supported by JST-CREST, and a Grant-in-Aid for Scientific Research in Priority Areas "Molecular Theory for Real Systems" from the MEXT of Japan, the Next Generation Supercomputing Project, and the Molecular-Based New Computational Science Program, NINS. The part of the computations has been performed using Research Center for Computational Science, Okazaki, Japan.

References

[1] Carroll T X, Hahne J, Thomas T D, Sæthre L J, Berrah N, Bozek J, Kukk E **2000** *Phys. Rev. A* 61 042503
[2] Kivimäki A, Kempgens B, Maier K, Köppe H M, Piancastelli M N, Neeb M, Bradshaw A M **1997** *Phys. Rev. Lett.* 79 998
[3] Hahne J A, Carroll T X, Thomas T D **1998** *Phys. Rev. A* 57 4971
[4] Dobrodey N V, Köppel H, Cederbaum L S **1999** *Phys. Rev. A* 60 1988
[5] Ehara M, Tamaki R, Nakatsuji H, Lucchese R R, Söderstrom J, Tanaka T, Hoshino M, Kitajima M, Tanaka H, Fanis A D, Ueda K **2007** *Chem. Phys. Lett.* 438 14
[6] Hatamoto T, Matsumoto M, Liu X-J, Ueda K, Hoshino M, Nakagawa K, Tahara T, Tanaka H, Kuramoto K, Ehara M, Tamaki R, Nakatsuji H **2007** *J. Electron Spectrosc. Relat. Phenom.* 155 54
[7] Adachi J, Kosugi N, Yagishita A **2005** *J. Phys. B* 38 R127
[8] Adachi J, Kosugi N, Shigemasa E, Yagishita A **1997** *J. Chem. Phys.* 107 4919
[9] Tanaka T, Ueda K, Feifel R, Karlsson L, Tanaka H, Hoshino M, Kitajima M, Ehara M, Fukuda R, Tamaki R, Nakatsuji H **2007** *Chem. Phys. Lett.* 435 182
[10] Tanaka T, Hoshino M, Kato H, Ehara M, Yamada N, Fukuda R, Nakatsuji H, Tamenori Y, Harries J R, Prümper G, Tanaka H, Ueda K **2008** *Phys. Rev. A* 77 012709
[11] Ehara M, Nakatsuji H **2008** *Coll. Czech Chem. Commun.* 73 771

Journal of Physics: Conference Series **235** (2010) 012020 doi:10.1088/1742-6596/235/1/012020

[12] Yoshida H, Nobusada K, Okada K, Tanimoto S, Saito N, Fanis A D, Ueda K **2002** *Phys. Rev. Lett.* 88 083001
[13] Tanaka T, Makochekanwa C, Tanaka H, Kitajima M, Hoshino M, Tamenori Y, Kukk E, Liu X J, Prümper G, Ueda K **2005** *Phys. Rev. Lett.* 95 203002
[14] Nobusada K **2002** *J. Phys. B* 35 3055
[15] Nakatsuji H **1978** *Chem. Phys. Lett.* 59 362
[16] Nakatsuji H **1979** *Chem. Phys. Lett.* 67 329
[17] Ehara M, Hasegawa J, Nakatsuji H **2005** *SAC-CI Method Applied to Molecular Spectroscopy, in Theory and Applications of Computational Chemistry: The First 40 Years, A Volume of Technical and Historical Perspectives*, Elsevier
[18] Nakatsuji H **1991** *Chem. Phys. Lett.* 177 331
[19] Ehara M, Nakatsuji H **1998** *Chem. Phys. Lett.* 282 347
[20] Ehara M, Nakatsuji H **2010** *Development of SAC-CI general-R Method for Theoretical Fine Spectrosocpy*, in P. Carsky J P, J. Paldus (Ed.), *Recent Progress in Coupled Cluster Methods: Theory and Applications*, Springer, Heidelberg
[21] Sankari R, Ehara M, Nakatsuji H, Senba Y, Hosokawa K, Yoshida H, Fanis A D, Tamenori Y, Aksela S, Ueda K **2003** *Chem. Phys. Lett.* 380 647
[22] Kuramoto K, Ehara M, Nakatsuji H **2005** *J. Chem. Phys.* 122 014304
[23] Ueda K, Hoshino M, Tanaka T, Kitajima M, Tanaka H, Fanis A D, Tamenori Y, Ehara M, Oyagi F, Kuramoto K, Nakatsuji H **2005** *Phys. Rev. Lett.* 94 243004
[24] Matsumoto M, Ueda K, Kukk E, Yohisda H, Tahara T, Kitajima M, Tanaka H, Tamenori Y, Kuramoto K, Ehara M, Nakatsuji H **2006** *Chem. Phys. Lett.* 417 89
[25] Ehara M, Nakatsuji H, Matsumoto M, Hatamoto T, Liu X-J, Lischke T, Prümper G, Tanaka T, Makochekanwa C, Hoshino M, Tanaka H, Harries J R, Tamenori Y, Ueda K **2006** *J. Chem. Phys.* 124 124311
[26] Sankari R, Ehara M, Nakatsuji H, Fanis A D, Aksela H, Sorensen S L, Piancastelli M N, Kukk E, Ueda K **2006** *Chem. Phys. Lett.* 422 51
[27] Ehara M, Kuramoto K, Nakatsuji H, Hoshino M, Tanaka T, Kitajima M, Tanaka H, Tamenori Y, Fanis A D, Ueda K **2006** *J. Chem. Phys.* 125 114304
[28] Worth G A, Beck M H, Jackle A, Meyer H-D **2003** *THe MCTDH Package, Version 8.3, University Heidelberg,*, Heidelberg, Germany
[29] Teffo J L, Chedin A **1989** *J. Mol. Spectrosc.* 135 389

AUTHOR INDEX

Abukawa, T 74
Ahamed, Jamal Uddin 80
Avaldi, L 45
Bolognesi, P 45
Borodin, Andriy 101
Casagrande, E M Staicu 27
Cautero, G 45
Chao, S D 10
Chigira, K 74
Chohda, M 127
Cvejanovic, D 33
Czasch, A 145
Ding, D 109
Dörner, R 145
Ehara, Masahiro 150
Enta, Y 62
Feifel, R 145
Foucar, L 145
Fujikawa, Takashi 68
Fukuzawa, H 109, 145
Gejo, T 116
Goto, T 74
Higashiya, A 145
Higuchi, I 109, 137
Hori, Masafumi 20
Horikawa, Y 116
Horio, Takuya 101
Ikeda, Akihiro 137
Ishikawa, T 145
Ishiyama, Takahisa 68
Ito, Kouta 68
Iwayama, H 109, 145
Katano, Satoshi 20, 80
Kato, H 62
Kawai, Maki 20
Kawata, H 74
Kim, Yousoo 20
Kimura, H 145
Kishimoto, N 85, 101
Konishi, Takehisa 68
Kono, S 74
Kono, Yutaro 137
Kotaki, T 74
Kudo, S 85
Kühne, K-U 145
Kurka, M 145
Lahmam-Bennani, A 27
Lin, S H 10
Liu, X-J 109, 145
Mineo, H 10
Moise, A 45
Motomura, K 145
Murakami, H 145
Nagaoka, Shin-Ichi 137
Nagasono, M 145
Nagaya, K 109, 145
Naja, A 27
Nakazawa, H 62
Napier, S 33
Neusser, H J 10
Ohashi, H 116, 145
Ohno, K 85, 101
O'Keeffe, P 45
Ooyama, K 74
Ouchi, T 109, 137
Oura, M 116
Ovcharenko, E 45
Patidar, Vinod 93
Piwinski, M 33
Pravica, L 33, 45
Purohit, G 93
Richter, R 45
Rudenko, A 145
Saito, N 109, 145
Sakai, K 109
Sakisaka, Y 62
Samarin, S 33
Sanbongi, Tomonori 80
Sato, S 62
Sawabe, A 74
Schlag, E W 10
Selzle, H L 10
Senba, Y 116
Sergo, R 45
Shin, S 116
Stebel, L 45
Sud, K K 93
Sugishima, A 145
Suzuki, Isao H 137
Tabayashi, K 127
Takahashi, O 116, 127, 137
Tamenori, Y 109, 137
Taniguchi, M 127
Teranishi, Y 10
Togashi, T 145
Tokushima, T 116
Trenary, Michael 20
Tsutsumi, Y 127
Ueda, K 109, 137, 145, 150
Uehara, Yoichi 80
Ullrich, J 145
Ushioda, Sukekatsu 80
Wada, Masahiko 68
Wang, Yi-Hsieh 10
Williams, J F 33
Woodruff, D P 1
Yabashi, M 145
Yamanaka, T 127
Yamazaki, Masakazu 101
Yao, M 109, 145

AUTHOR INDEX

Yoshida, H **127**
Zhang, D **109**
Zhu, Lin-Fan **53**

Institute of Physics
Dirac House, Temple Back
Bristol BS1 6BE UK

CURRAN ASSOCIATES INC
proceedings.com